VANGUARD SERIES

EDITOR: MARTIN WINDROW

GERMAN LIGHT PANZERS 1932-1942

Text by
BRYAN PERRETT

Colour plates by
TERRY HADLER

OSPREY PUBLISHING LONDON

Published in 1983 by
Osprey Publishing Ltd
Member company of the George Philip Group
59 Grosvenor Street, London, W1X 9DA
© Copyright 1983 Osprey Publishing Ltd
Reprinted 1985, 1987, 1988, 1989

This book is copyrighted under the Berne Convention.
All rights reserved. Apart from any fair dealing for the
purpose of private study, research, criticism or review,
as permitted under the Copyright Act, 1956, no part
of this publication may be reproduced, stored in a
retrieval system, or transmitted in any form or by any
means, electronic, electrical, chemical, mechanical,
optical, photocopying, recording or otherwise, without
the prior permission of the copyright owner. Enquiries
should be addressed to the Publishers.

British Library Cataloguing in Publication Data

Perrett, Bryan
    The German light Panzers 1932–1942.—
    (Vanguard series; 33)
    1. Germany. *Heer. Panzerjager*—History
    I. Title   II. Series
    358'.18   UG446.5

    ISBN 0-85045-483-2

Filmset in Great Britain
Printed in Hong Kong

Acknowledgements
Author and publishers are grateful to Bruce Culver
and Steven Zaloga for their assistance in assembling
illustrations for this book.

PzKpfw I Ausf.A tanks manoeuvre during a training exercise, probably in summer 1935 at Munster Lager. A hard-edged pattern, probably of brown over grey, is painted over the hulls and turrets; very early examples sometimes displayed a three-colour scheme in grey, green and brown. The only markings visible here are a crude black 'club' playing-card symbol on the rear trackguard flaps, presumably indicating the sub-unit. (RAC Tank Museum)

The Background

Hindsight has been described as a mirror which enables ordinary men to be wise after the event. This may well be true, but it is also true that hindsight frequently offers a distorted perspective, so that the contemporary view of events and equipment held two generations ago is rather different from that held today. Thus, because of the spectacular nature of its victories from 1939 to 1941, it is tempting to regard Hitler's Panzerwaffe as a superbly equipped cutting edge for the rest of the army. In reality, the Panzerwaffe excelled only in technique, and was very badly equipped. Something of the delays experienced in its equipment programme have been mentioned in Vanguard No 16 *The Panzerkampfwagen III* and Vanguard No 18 *The Panzerkampfwagen IV*, and since these vehicles were in critically short supply it was upon a mass of light tanks that the Panzerwaffe relied to fill out the ranks of its under-strength divisions during its high years. This amply justifies a study of the latter, if only on the grounds that their apparent lack of potential makes their achievement the more remarkable; nor, indeed, are they entirely without points of technical interest.

In terms of general tank design, the outbreak of war in 1939 found Germany behind the Soviet Union and France, almost level-pegging with Great Britain, slightly ahead of Japan, and with a decisive lead over Italy. The United States was not yet a contender in the race, but was well advanced in the designs which led to the Lee and the Stuart (see Vanguard No 6 *The Lee/Grant Tanks in British Service* and Vanguard No 17 *The Stuart Light Tank Series*). For this the restrictive clauses of the Treaty of Versailles, under which Germany was forbidden tracked AFVs, are sometimes blamed, but the fact is that German tank designers were at work long before Hitler repudiated those clauses in 1935. During the 1920s a secret experimental station was established jointly with the Red Army deep inside Soviet Russia, and much technical data had also been gained from work carried out discreetly in Sweden. Again, German military attachés throughout the world were fully conversant with the latest developments in tank design, details of which were contained in their reports; nor should the national reputation for expertise in heavy engineering be forgotten.

Crews and vehicles lined up for inspection at the end of a demonstration or exercise. The original print shows the playing-card trackguard symbols; and note white disc, presumably an exercise recognition device, painted at the right of the rear hull plate and on the turret hatch. The nearest tank at the right seems to display a three-colour pattern. (RAC Tank Museum)

The reasons, therefore, must be sought elsewhere. First, and most important, was the fact that the newly formed Armoured Corps certainly did not envisage becoming involved in a major war as early as 1939, and did not believe that it would be prepared for such an undertaking until approximately 1943. Secondly, it was entirely reasonable that it should order for itself a simple, inexpensive machine, the PzKpfw I, with which to carry out preliminary crew training *en masse*. Thirdly, while the basic concepts of its two main battle tanks, subsequently known as the PzKpfw III and IV, were essentially sound, it seems that the army's procurement department, the Heereswaffenamt, seriously under-estimated the time required for these to reach full scale standardised production; this being the case, the Panzerwaffe felt compelled to order a stop-gap vehicle, the PzKpfw II, which was also capable of reconnaissance but which was simply an extension of the light tank theme.

It was certainly not compatible with the National Socialist ethos to make good the shortfall with vehicles purchased abroad, even if a trading partner could have been found. On the other hand, the rape of Czechoslovakia was accompanied by the acquisition of the Czechs' own tank fleet and domestic manufacturing resources; in such circumstances it would have been extremely foolish to do other than return pride to one's pocket, since the Czech vehicles, known in German service as the PzKpfw 35(t) and PzKpfw 38(t)—(t) for Tschechisch, 'Czech'—were similarly armed to the PzKpfw III and could be substituted for it.

The Tanks

Panzerkampfwagen I (SdKfz 101)

The traditional policy of the Heereswaffenamt was to issue basic specifications for a project to a number of civilian manufacturers and then choose the best design submitted, and in 1932 it despatched competitive contracts for a light tracked fighting vehicle which, for the moment, was known simply as the Landwirtschaftlicher Schlepper (LaS), or Industrial Tractor.

The Krupp entry, incorporating experience gained jointly with the Swedish Landsverk company during examination of a British Carden Loyd chassis, was selected; and the initial manufacturing contract was given to the Henschel organisation under the interim designation of 'I A LaS Krupp'. The first prototypes were delivered in December 1933 and quantity production began in July 1934, the vehicle's official service title being **PzKpfw (MG) I Ausführung A** once the need for subterfuge had disappeared.

As might be expected in what was intended essentially to be a training machine, the design was very simple. The suspension consisted of four bogie wheels and a trailing idler, braced by an external beam and secured to the hull by bolts and quarter-elliptical leaf springs; track adjustment was obtained by altering the position of the idler. The power unit was a four-cylinder horizontally opposed air-cooled Krupp petrol engine which produced 57hp at 2,500rpm, fitted with one Solex downdraft carburettor for each bank, and an acceleration pump. (A few models were fitted experimentally with an air-cooled diesel engine, but this produced only 45hp at 2,200rpm, insufficient for the vehicle's requirements.) Fuel capacity was 20 gallons, housed in two tanks mounted in the rear corners of the driving compartment.

From the engine the drive passed through a two-plate dry clutch to the gearbox, which provided one reverse and five forward gears, and thence across the front of the vehicle to the drive sprockets. Steering was effected by clutch and brake, cooling being supplied by a small fan. The driver controlled direction by means of steering levers, each of which had two hand-grips—one for normal steering and the other with a thumb plunger to act as a parking brake; no handbrake was fitted. The instrument panel contained an oil temperature gauge; a revolution counter marked from 0–3,000rpm with a danger zone above 2,500rpm; and a speedometer marked from 0–50kph. Maximum recommended speed in gears was: 1st–5kph; 2nd–11kph; 3rd–20kph; 4th–32kph; 5th–42kph.

The second crew member, the commander/gunner, was housed in a small turret mounted to the right of the vehicle's centre line. All-round traverse was available for the turret's two 7.92mm MG13 machine guns, which were mounted in tandem but capable of independent fire; the gun on the left was fired from a trigger on the elevating handwheel to the commander's left, that on the right being fired from a trigger on the traverse handwheel to his right. Maximum elevation obtainable was +18° and maximum depression −12°. A clutch was incorporated in the elevation handwheel which gave 'Operate' and 'Free' positions, and the mantlet itself could be locked in the horizontal. Ammunition was contained in 25-round clips, stowed as follows:

1 bin in turret containing 8 clips
1 bin in hull containing 8 clips
1 bin in hull containing 20 clips
1 bin in hull containing 6 clips
1 bin in hull containing 19 clips

The commander's seat was suspended from the turret and rotated with it, although the floor of the fighting compartment remained static. His communication with the driver was by means of voice tube.

Armour thickness was 13mm all round, proof against small arms ammunition but little else. Trials with a captured example revealed that even sustained machine gun fire was capable of jamming the turret ring and mantlet, while the air intake was considered to be highly vulnerable to grenade attack.

Those vehicles fitted with radio mounted a collapsible aerial on the offside of the hull, operated by a handle inside the fighting compartment. When the turret was traversed a cam acting

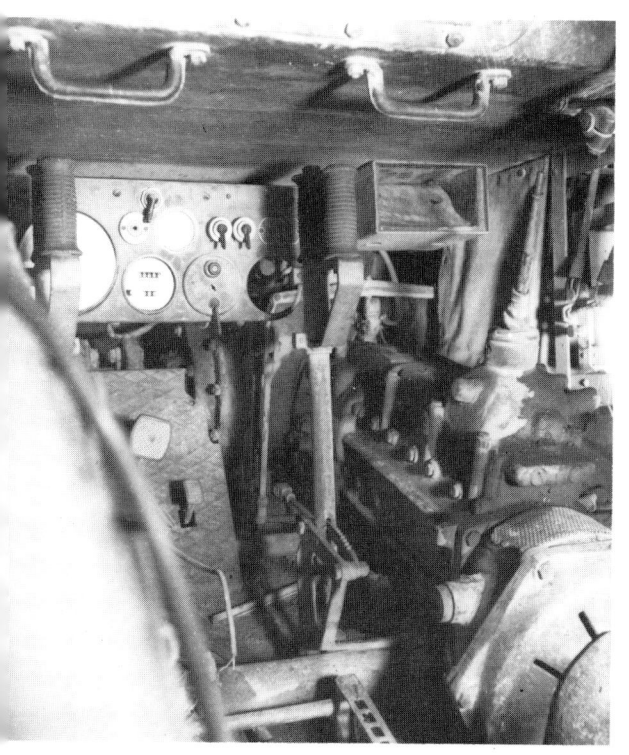

Driver's controls and instrument panel of a PzKpfw I Ausf.A (RAC Tank Museum)

The Kleiner Panzerbefehlswagen (SdKfz 265) on the Ausf.B chassis, photographed at an Armeeoberkommando—AOK—in 1941. (US National Archives)

on the turret ring automatically lowered the aerial to prevent it fouling the guns.

No sooner had the Ausf.A entered service than it was found to be badly underpowered for cross-country work. Some redesign was clearly necessary, and this resulted in the **Ausführung B**, driven by a six-cylinder water-cooled Maybach NL 38 TR engine and fitted with an improved transmission. Being larger, the Maybach unit occupied more space than was available, this deficiency being remedied by extending the engine compartment to the rear. This, in turn, meant a longer suspension incorporating five bogie wheels, and because of this the idler wheel was raised above ground level to facilitate steering. In all other respects the Ausf.B was almost identical to its predecessor. Some 300 Ausf.A versions were built, and approximately 1,500 Ausf.B.

In September 1939 it was decided to develop the design into a light fighting vehicle which could undertake the reconnaissance role and also provide support for air-landing operations. The work was jointly carried out by Krauss-Maffei and Daimler Benz and resulted in the **Ausführung C**. This vehicle, armed with a 20mm cannon and a co-axial machine gun, had a five-wheel interleaved suspension and was powered by a Maybach 150hp engine. While it reached the prototype stage it does not seem to have passed beyond this.

A further development was the **PzKpfw I nA verst** (neue Ausführung, verstarkt—'new model, up-armoured'), which is sometimes referred to as the **Ausführung D or F**. Requested in December 1939, this vehicle lay outside the basic philosophy of the Panzerwaffe in that it was intended for infantry support. The project was again awarded to the team of Krauss-Maffei and Daimler Benz, which employed the same chassis and engine as the Ausf.C. Armament consisted of two MG34 machine guns mounted in tandem, and armour thickness was 80mm, giving a total weight of 18 tons. Maximum speed was 15mph, quite acceptable for the role envisaged and comparable to that of British Infantry Tanks. The vehicle began entering service in June 1940, and 30 are known to have been delivered before the order was cancelled, although details of their eventual use remain obscure.

PzKpfw I Derivatives

Probably the best known derivative of the PzKpfw I was the **Kleiner Panzerbefehlswagen** ('small armoured command vehicle'), which employed both the Ausf A and B chassis. This vehicle was fitted with a fixed superstructure in place of a turret, and was armed with a ball-mounted machine gun for local defence. Two radios were carried, the Fu2 and the Fu6, and the vehicle was manned by a crew of three. About 200 of these conversions were produced by Daimler Benz between 1936 and 1938, emphasising the fact that from the outset tactical control from well forward was an essential element in the Panzer-waffe's operational technique.

Ausf.B chassis were extensively used as mountings for self-propelled guns, the conversions being undertaken by the Alkett organisation of Berlin-Spandau in 1939. These included the Wehrmacht's first tracked tank destroyer, the **PaK 4.7cm(t) Sfl auf PzKpfw I Ausf.B** which, as its title implies, was fitted with a Czech anti-tank gun. The mounting for this 43.4-calibre-length weapon was protected on three sides by armour plate and provided a limited traverse of 15°. A total of 86 rounds of ammunition could be stowed, and a crew of three was carried. These vehicles remained in service on the Eastern Front until the winter of 1941–42, and were also encountered in North Africa.

The Type 33 150mm Heavy Infantry Gun,

complete with carriage and wheels, was also mounted on the Ausf.B chassis, the 38 conversions made being designated **150mm sIG 33 auf PzKpfw I Ausf.B**. The mounting was protected on three sides by 10mm armoured shields which gave the vehicle an ungainly appearance, increasing its height to over ten feet; together, the shields and weapon combined to bring the overall weight to 8.5 tons, which grossly overloaded the chassis and curtailed mobility. These conversions were manned by crews of four, and saw active service during the Polish and French campaigns.

Turretless, open-topped chassis were used for driver training and as supply carriers, and some Ausf.B chassis were issued to tank recovery and repair units as tractor units.

The **Ladungsleger I** (Explosive Charge Layer I) was an interesting venture into the field of assault engineering. The equipment consisted of a pair of telescopic arms mounted on a framework above the engine deck on an Ausf.B. These could be extended to deposit a 75kg charge on to the roof of a pillbox or similar fortification, the charge being detonated by remote control once the vehicle had moved out of the danger radius. A prototype was completed towards the end of 1940, but the idea was not widely developed.

Panzerkampfwagen II (SdKfz 121)

As already mentioned, the PzKpfw II was intended to be a stop-gap machine pending the arrival of the PzKpfw III and IV, and its main armament specification of one 20mm cannon could still be regarded as adequate—but only just—in July 1934, when the Heereswaffenamt put the project out to competitive tender under the cover name of 'LaS 100'.

After the various prototypes had been evaluated, that submitted by Maschinen-Fabrik Augsburg-Nürnberg (MAN) was selected for further development. Layout was almost identical to that of the PzKpfw I, and indeed the first four production models, **Ausf.a1, a2, a3 and b**, possessed very similar characteristics, employing leaf springs and an external beam for the suspension of six small bogies, arranged in three pairs of two. The Ausf.a1, a2 and a3 were all driven by a Maybach six-cylinder HL 57 engine, which produced 130hp at 2,600rpm. These three models, of which 25, 25 and 50 respectively were manufactured, reflected

A Ladungsleger I engineer tank, with 75kg explosive charge supported by a frame above the rear superstructure. Small numbers of these were used by the Panzer-Pionier units of the original ten Panzer-Divisionen in 1940 and 1941, but they proved too vulnerable under combat conditions. (RAC Tank Museum)

7

PzKpfw I Ausf.A of the 1st Company, 1st Battalion of Combat Tanks, Spanish Nationalist Army, 1938; compare markings with Plate A2. In the background, a captured Republican T-26 Model 1933; this Soviet type was greatly superior to the PzKpfw I, and captured examples were pressed into Nationalist service whenever possible. (via Steven Zaloga)

minor but progressive improvements, notably to the cooling system, but lacked final drive reduction gears and were soon shown to be under-powered. These difficulties were resolved by fitting the Ausf.b with a Maybach six-cylinder HL 62 engine, which had an output of 140hp, and modifying the final drive to incorporate a reduction train. One hundred vehicles of this type were built.

The fourth production model, **Ausf.c**, appeared in 1937 and saw an increase in frontal armour thickness from 14.5mm to 30mm. It also saw the introduction of the five-wheel quarter-elliptical leaf spring suspension by which the PzKpfw II is generally remembered.

At this point it was decided to accelerate production and MAN were joined by the Henschel, Famo, MIAG and Wegmann organisations in building the numbers required. Until now the bow of the vehicle consisted of a one-piece rounded casting, but this was replaced on the **Ausf.A** and all subsequent models by a square joint-welded combination of flat bow and glacis plates. The turret of the **Ausf.B** was fitted with a shallow cupola, as was that of the **Ausf.C**, these two models being identical in appearance although the latter incorporated several interior improvements.

In 1938 Daimler Benz were contracted to produce a Schnellkampfwagen ('fast fighting vehicle') version of the PzKpfw II with which to equip the Light Divisions. This was achieved by replacing the standard suspension with a torsion bar system mounting four large Christie-type roadwheels, the return rollers being dispensed with. In this form the vehicle was capable of a maximum speed of 55kph. Two almost indistinguishable models, **Ausf.D** and **E**, were built, the total production run being 250 vehicles. However, due to internal army politics, very few actually reached the Light Divisions and the majority served in Panzer-Regiment 8 of Panzer-Brigade 4 which, as we shall see, had a very different role.

By 1939 the Panzerwaffe possessed a stock of

1,266 PzKpfw IIs and production was allowed to tail off, only 15 being completed in that year and nine in 1940. Hitler's decision to double the number of Panzer divisions immediately led to a resumption in order to flesh out the establishment of the new formations, the vehicle's by now deficient firepower and protection having been masked to some extent by the easy victories in Poland and France. The **Ausf.F** began entering service at the beginning of 1941, protected by 35mm frontal and 20mm side armour; the effect of this modest attempt at up-armouring was an increase in vehicle weight to 9.35 tons, purchased at the cost of a reduction in speed by 10kph to 40kph. The series continued with the **Ausf.G** and **J**, which were distinguished from the Ausf.F by the addition of a stowage bin to the turret rear. Production did not finally cease until the beginning of 1944.

The PzKpfw II's turret was offset to the left of the tank's centre-line, and the vehicle was manned by a crew of three: commander/gunner, loader/operator and driver. The line of drive passed from the engine through a plate clutch to a crash gearbox which provided one reverse and six forward gears, and thence across the vehicle's front to the drive sprockets. Maximum recommended speed in gears was: 1st–5kph; 2nd–10kph; 3rd–12kph; 4th–21kph; 5th–30kph; 6th–40kph. Steering was effected by levers operating a clutch and brake mechanism. Driver's instruments included a speedometer; a revolution counter calibrated from 1,000 to 3,200rpm, marked in red above 2,600rpm; a water temperature gauge; and an oil pressure gauge. An electric starter was provided, but for low temperatures an inertia starter, operated through the vehicle's stern plate, was also fitted.

In addition to a 20mm cannon the PzKpfw II was also armed with a co-axial 7.92mm MG34 machine gun. The 20mm was fired from a trigger on the elevating handwheel to the commander's left, and the MG34 from a trigger on the traverse handwheel to his right, the latter giving four degrees of traverse per full turn. The traverse mechanism operated through a dog clutch controlled by three levers marked as follows:

EIN ('geared'): Traverse gear engaged, traverse lock disengaged

AUS ('free'): Traverse gear disengaged, traverse lock disengaged

EST ('locked'): Traverse gear disengaged, traverse lock engaged

In the 'free' position it was possible to turn the turret by hand quite quickly, using two handles on the turret ring. Zero marks on the hull and turret coincided when the latter was in the 12 o'clock, i.e. fully forward, position.

Ammunition was stowed in the following manner:

20mm—In ten-round magazines. On the nearside (left) wall of the fighting compartment —four magazines. On the offside wall of the fighting compartment—eight magazines. Under the turret ring, offside front of fighting compartment—six magazines. Total: 18 magazines containing 180 rounds.

7.92mm—In belts contained in protective leather bags. On a rail low on the nearside wall of the nearside of the fighting compartment—11 belt bags. On a rail above this—five belt bags. On a rail fitted to the forward nearside of the hull—one belt bag. Total: 17 belt bags containing 2,550 rounds.

In dusty climates the mouth of the belt bag was usually stuffed with a wad of grease-proof paper, keeping the rounds free from grit which could cause jamming in the weapon's feed mechanism. Examination of belts stowed aboard PzKpfw IIs

An early Kleiner Panzerbefehlswagen, without cupola, in service with the Legion Condor—note St. Andrew's cross, and Nationalist flag marking on hull front. The machine gun lacks the later ball mount. (via Steven Zaloga)

captured in North Africa revealed a mix of half-and-half alternate AP/AP tracer, or sometimes 100 per cent AP tracer, but often belts had obviously been filled hastily with whatever was to hand, no thought being given to sequence.

Two types of gunsight were employed: an open sight graduated from 200 to 800 metres with 200-metre markings; and a telescopic TZF 4/38 sight with a magnification of 2.5 and a field of 25°, graduated from 0 to 1,200 metres with 200-metre markings.

Attached to the tank's stern plate was a rack of five smoke grenades which could be released by pull-wires from inside the fighting compartment, enabling the vehicle to reverse into its own smokescreen.

The PzKpfw II remained under-armoured throughout its active service career and, as in the case of the PzKpfw I, its turret ring was vulnerable to sustained machine gun fire. A small point of interest on later models was an aluminium dummy visor fitted beside the driver's real visor, clearly in the hope that it would attract a portion of the enemy's fire.

This photo shows Legion Condor PzKpfw I Ausf.A tanks in action in Spain. (Imp. War Mus.)

Normally, three radio sets—two receivers and one transmitter—were carried, the aerial being mounted at the nearside rear of the fighting compartment. Internal communication between the commander and driver was by means of voice tube.

In parallel with the standard PzKpfw II gun tank, a reconnaissance version had also been under development since 1938. Using experience gained from various experimental projects MAN produced a prototype in April 1942, and the vehicle began entering service the following year under the designation **PzKpfw II (SdKfz 123) Ausf.L**, later amended to **Panzerspahwagen II (20mm KwK 38) Luchs** ('Lynx').

The Lynx weighed 11.8 tons, was powered by a Maybach HL 66P 180hp engine, and could achieve a maximum speed of 60kph. The suspension consisted of five large interleaved bogie units carried on torsion bars. Frontal armour was 30mm thick, and the first 100 production models were armed similarly to the PzKpfw II Ausf.A–J; the remaining 31 vehicles, built before manufacture ceased in May 1943, were armed with the same 50mm L/60 weapon mounted by the PzKpfw III Ausf.J–M, producing an almost unworkable

combination of too much gun for too little turret. Much emphasis in the design stage had been placed on medium and short wave radio equipment and this, together with an additional operator, resulted in an impossibly cramped interior.

Technically the Lynx was an Aufklärunspanzer or armoured reconnaissance vehicle, and it was used exclusively by armoured reconnaissance battalions. Plans for a more heavily armoured version mounting a 50mm gun and co-axial MG42 machine gun were completed in 1942, but the project was cancelled—although the turret was later fitted to the eight-wheeled Puma armoured car (see Vanguard No 25, *German Armoured Cars & Reconnaissance Half-Tracks*). Had this vehicle been built, it would have been known as the 'Leopard'.

PzKpfw II Derivatives

During the preparations for Operation 'Sea Lion', Hitler's planned invasion of the south coast of England, it was decided to form an amphibious tank battalion equipped with PzKpfw IIs. Unlike the amphibious versions of the PzKpfw III and IV, however, the **Schwimmpanzer II** would not have to drive along the sea bed to reach the shore; instead, as its name implies, it would swim ashore from its parent vessel, using a kit of floatation tanks attached to the return rollers, and powered by a propeller driven by an extension shaft from the engine. The turret ring was sealed by an inflatable rubber ring. Again, unlike the 'diving' versions of the medium tanks, the Schwimmpanzer II could use its guns during the landing, and was expected to.

The unit which would have manned these vehicles had the operation taken place was known as 'Panzerabteilung A', recruited from volunteers drawn from Panzer-Regiment 2. In due course this and other volunteer battalions were formed into Panzer-Regiment 18, which carried out an amphibious crossing of the River Bug during Operation '*Barbarossa*'. It is just possible that a number of Schwimmpanzer IIs may have participated in this, although the bulky floatation equipment made the vehicle more suitable for landing across a beach than for a river crossing in which it would have to negotiate difficult banks.

Interesting view of one of the several PzKpfw Is modified to mount a 20mm cannon after the battle of Sesena in Spain showed up their poor armament when faced by Soviet types. (via Steven Zaloga)

11

The PzKpfw I's armour was barely sufficient to protect it from machine gun fire. (Above), a tank with the hull front completely caved in by a direct hit from a Polish 75mm field gun during September 1939: 53 of the 89 PzKpfw Is lost in Poland were total combat losses. (via Steven Zaloga) (Right), a PzKpfw I Ausf.B of 4.Pz-Div. in France, 1940, also completely destroyed by an artillery hit. (US Nat. Archives)

Another imaginative project was the conversion of 95 Ausf.D and E chassis to the flamethrower role. The conversion was generally known as the **Flammpanzer II** but was also referred to as the 'Flamingo'; its official designation was **PzKpfw II(F) (SdKfz 122)**. Two flame guns, each with a traverse of 180°, were mounted well forward, and sufficient fuel was carried for 80 two- to three-second flamings. The range of the flame guns was limited to 35 metres, and the vehicle was clearly vulnerable at such close quarters, particularly as the original turret had been replaced by a smaller one mounting a single machine gun. Photographic evidence suggests that in order to remedy this deficiency some vehicles were fitted with a battery of grenade projectors above their engine decks. Because of the space occupied by the flame equipment and fuel the crew was reduced to commander and driver. The Flammpanzer II served in specialist battalions which were employed at the discretion of senior commanders.

By 1942 it was painfully obvious that the PzKpfw II's usefulness as a gun tank was at an end, and large numbers of chassis were turned over for conversion to other uses. It was, in fact, very fortunate for the Wehrmacht that they were available, since they served as suitable mountings for its first generation of tank destroyers, the 75mm L/46 PaK 40/2 anti-tank gun being fitted to Ausf.A, B, C and F chassis and the captured Russian Model 36 76.2mm anti-tank gun to Ausf.D and E chassis. Both these conversions were known as **Marder II**, and are more fully discussed in Vanguard No 12 *Sturmartillerie and Panzerjäger*.

The 150mm s.I.G. 33 was also mounted on Ausf.A, B, C and F chassis, the conversion being lower and generally more businesslike than that based on the PzKpfw I. Weight and space considerations led to later models being based on an extended chassis. The vehicle weighed approximately 12 tons and was manned by a crew of five.

Best known of the self-propelled mountings based on the PzKpfw II was the 105mm howitzer

Wespe ('Wasp'), the official title of which was **le FH 18/2 auf Fahrgestell PzKpfw II Sf**; for reasons best understood by himself, Hitler decided that the name Wespe should be dropped in February 1944. This vehicle began entering service in 1942 and equipped the light batteries of Panzerartillerie regiments. It was built in substantial numbers, using most PzKpfw II chassis with the exception of Ausf.D and E. The fighting compartment, surrounded by an open-topped superstructure of armour plate, was located at the rear. A total of 32 rounds of ammunition were stowed, and the vehicle was manned by a crew of five. The L/28 howitzer was fitted with a muzzle brake and could be elevated to $+42°$ and depressed to $-5°$; $17°$ of traverse were available either side of the centre line. A machine gun was provided for local defence.

In spite of its handy appearance, the Wasp was not universally popular with professional artillerymen, who considered it to be cramped, too high at 7ft. 8ins., and too poorly protected for its forward fire support role. Within these limitations, however, the vehicle performed well. A further area of criticism, namely that it carried too little ammunition, was partly remedied by the production of an ammunition carrier version, in which the howitzer was omitted, capable of stowing 90 rounds.

Panzerkampfwagen 35(t)

During the 1930s the Czech armaments industry was one of the most highly developed in Europe, and its exports made a substantial contribution to the wealth of the country. In 1935 the two fighting vehicle manufacturers, Ceskomoravska Kolben Danek (CKD) and Skoda, decided jointly to produce a light tank the designation of which was LTvz.35, although in German service it was to be known as PzKpfw 35(t). Unlike the German light tanks, the vehicle's armament was comparable to that of the PzKpfw III, consisting of a 37mm gun and co-axial 7.92mm machine gun, with a further machine gun in the front plate; the 25mm frontal armour, too, exceeded by a wide margin that planned for the first models of the PzKpfw III and IV, yet in spite of this the designers had succeeded in producing a compact vehicle weighing only 10.5 tons and capable of a maximum speed of 35kph.

The PzKpfw 35(t) was crewed by a commander/gunner, loader, driver and hull gunner/operator. The commander was located on the left of the turret beneath a rudimentary cupola incorporating four episcopes and a one-piece circular hatch. The loader occupied the opposite side of the turret. In the driving compartment the hull gunner was seated on the left and the driver on the right.

Much priceless internal space had been conserved by the designers' decision to employ a rear drive sprocket, which dispensed with the need for bulky transmission units inside the driving and fighting compartments. The vehicle was powered by a four-cylinder 120hp Skoda T/11 engine, and the gearbox provided six forward and six reverse gears. A planetary steering system was employed, both steering and gear changing being assisted by compressed air. This made the PzKpfw 35(t) a pleasure to drive in normal conditions, although the low temperatures encountered during the Russian winter adversely affected the pneumatic equipment.

The suspension consisted of two sets of double bogie wheel pairs, each set hanging from a large leaf spring bolted to the hull side. This system equalised wear to such an extent that daily road

The PzKpfw Is and IIs of 4.Pz-Div. spread out across the fields of Belgium during the advance of Hoepner's XVI Panzer-Korps to the Gembloux Gap in May 1940. (US Nat. Archives)

marches in excess of 100 miles could be achieved, provided the speed remained moderate, and in some cases a quite remarkable track life of 5,000 miles was recorded. A further point of interest concerning the suspension was a small track tensioning wheel located between the forward bogie and the front idler.

The hull and turret were constructed from riveted plates, and in this respect the vehicle fell below the specification of its German contemporaries. It was not widely appreciated at the time that the effect upon a rivet head of a high velocity round was to sheer off the shank which, having absorbed much of the strike's terminal velocity, then flew round the inside of the hull killing and maiming.

The vehicle's main armament was an adaption of the Skoda A3 37mm anti-tank gun, fitted with a very simple perforated muzzle brake, and had a semi-automatic falling block type breech. The recoil cylinder projected some way beyond the mantlet and was protected by a prominent armoured sleeve. The gun fired AP shot at a muzzle velocity of 675 metres per second and was capable of penetrating 30mm armour at approximately 600 yards. The two 7.92mm machine guns were belt-fed air-cooled weapons developed by Ceska Zbrojovka of Brno; interestingly, a licence to manufacture this gun was sold to the BSA organisation's Birmingham plant in 1937 and it evolved into the BESA machine gun, which was mounted by British tanks throughout World War II and beyond. Stowage consisted of 72 rounds of main and 1,800 rounds of secondary armament ammunition.

In many respects the PzKpfw 35(t) was an advanced design for its day, and it was always popular with its crews. Its best known users were 6.Panzer-Division, but it also saw service with the Rumanian 1st Royal Armoured Division, with

One of the PzKpfw I Ausf.As of 4.Pz-Div. during the campaign in the West, spring 1940; the divisional sign can be made out on the hull front plate beneath the mantlet. The Ausf.A's shorter suspension is very evident in this shot. (US Nat. Archives)

the Slovak Fast Division, and in limited numbers with the Bulgarian Army. As far as the Panzerwaffe was concerned, its career as a gun tank ended with the winter of 1941–42, but in the hands of satellite troops it remained in action until the Stalingrad campaign.

Some standard versions of the tank were equipped as command vehicles (**Panzerbefehlswagen 35(t)**) for the 1939 campaign in Poland, extra radios being installed and a collapsible frame aerial erected over the engine decks. In general, however, the configuration of the vehicle was such that it was unsuitable for conversion to specialist roles, so that when chassis did become available through the obsolescence of the gun tank they were used simply as towing vehicles. Two versions, both turretless, existed: one, the **Zugkraftwagen 35(t)**, was used by vehicle recovery sections; the other, known as the **Morserzugmittel 35(t)**, served in the artillery's heavy mortar batteries.

Panzerkampfwagen 38(t)

Notwithstanding its ultimate popularity, the PzKpfw 35(t) suffered an initial unreliability which caused adverse comment within the Czech Army. This reflected a somewhat impatient lack of understanding that a certain amount of modification to detail is inevitably required to perfect a design even after it has entered service. Nonetheless, in 1937 the Staff requested competitive tenders for a new battle tank from Skoda and CKD (which became known as Bohmisch-Mahrische Maschinenfabrik AG after the German takeover) after the latter's TNHP design was accepted after trials.

In German service the TNHP was designated PzKpfw 38(t). The vehicle was similarly armed and armoured to the PzKpfw 35(t) and was driven by a 125hp Praga EPA six-cylinder water-cooled engine which produced a maximum speed of 42kph; on later models the engine was fitted with twin carburettors which raised the output to 150hp and the top speed to 48kph. A front drive sprocket was employed, power reaching this through a five-forward/one-reverse Praga-Wilson gearbox. The suspension consisted of four large Christie-type roadwheels per side, hanging in pairs from horizontal leaf springs bolted to the hull. The

Kleiner Panzerbefehlswagen of 4.Pz-Div. used as an ambulance vehicle in France, 1940; note red cross markings on front and sides, and divisional insignia just visible above the visor. Other ambulance conversions of the PzKpfw I had cut-down, open-topped superstructure. (US Nat. Archives)

steering system was similar to that installed on the PzKpfw 35(t) but lacked pneumatic assistance. Fuel capacity was 40 gallons, contained in two double-skin petrol tanks mounted on either side of the engine compartment. An electric starter was provided, as was an inertia starter operated from the rear of the vehicle; in emergencies the engine could be started manually from inside the fighting compartment by a device located on the flywheel housing. The position of the rear idler, which controlled track tension, could be adjusted through hatches in the stern plate.

Hull and turret construction was again of riveted plate, although rather less rivets were used than in the PzKpfw 35(t). Internally, the position of the four crew members was identical. The commander's cupola contained four episcopes, and immediately forward of this was a panoramic periscope.

The tank's main armament was the improved Skoda A7 37mm, known in German service as the KwK 37(t). It was a semi-automatic falling block weapon which fired AP shot at a muzzle velocity of 750 metres per second and could penetrate 32mm armour at 1,100 metres; an HE round was

The Panzerjäger IB, photographed here during training in France in 1941; the insignia of Pz-Jäg-Abt.521 is painted in white on the shield. The mottled effect is leaf-shadow. Note helmets slung over the front of the shield. (US Nat. Archives)

also developed for this gun. The piece was slightly breech heavy and the recoil cylinder projected a little way beyond the mantlet. Elevation was obtained either by an inconvenient horizontal handwheel to the gunner's right, or by means of a curved crutch which fitted his right shoulder; the latter system could be engaged by removing a worm from the elevating arc, but was extremely tiring. The traverse handwheel was badly sited on the gunner's left, requiring awkward bending of the wrist. Both the main and co-axial armament could be fired from a trigger on the elevating handwheel, which also incorporated a safety button; a gun selector lever was located under the 37mm giving four positions, viz: main, main and co-ax, safe, and co-ax. The sighting telescope included both 37mm and machine gun range scales, and in the event of it being rendered inoperable a small open sight for emergency use had been drilled through the armour below. The co-axial mounting could be unlocked and the weapon used independently of the main armament with 10° of traverse, 20° of elevation and 10° of depression. The bow machine gun could be fired either by the hull gunner or the driver, the latter having an additional trigger fitted to his left tiller bar. Ninety rounds of 37mm and 2,700 rounds of machine gun ammunition were stowed, the majority in the bulge at the turret rear. The radio was mounted on the hull wall to the left of the bow gunner.

The PzKpfw 38(t) ran to several models of which the **Ausf.A** was the standard version taken over on the occupation of Czechoslovakia. The majority differed only in minor detail, such as the installation of a smoke grenade rack at the rear (**Ausf.B**); but with **Ausf.E** the armour thickness was doubled by bolting 25mm plates to the front of the vehicle and 15mm plates to the side. The **Ausf.G** was the last production model before chassis were built exclusively for self-propelled mountings, while **Ausf.S** was a version built for Sweden but confiscated by the German Army before delivery could be made.

Altogether 1,414 PzKpfw 38(t)s were built. Its reliability and ease of maintenance made it a very attractive acquisition, so that in addition to widespread use by the German Army it was also exported to Iran, Rumania, Bulgaria, Hungary, Sweden, Switzerland and Peru. Even the Royal Armoured Corps, in the throes of re-armament and desperately short of equipment, was sufficiently interested to arrange for CKD to deliver a demonstration model to the Gunnery School at Lulworth on 23 March 1939, and firing trials took place the next day. The following extracts from the report prepared after the trials reflect the RAC's thoughts on crew practice in what was evaluated as a medium tank turret, and these, of course, coincided with those of the Panzerwaffe; they also reflect the professional user's view that in achieving their equation of mobility, firepower and protection tank designers frequently ignore the human element:

'On the whole the machine is almost equivalent to our cruiser tanks, but little experience or experiment has gone into the design of the fighting compartment and performance has been obtained at the expense of the crew and general 'fightability'.

'The tank was comfortable to ride in and judging by the way it took jumps there is very little chance of the crew being injured when travelling over unknown country. The stiffness of the suspension, however, set up a judder or wobble in the turret, making it impossible to lay the guns when travelling at over 5mph. Even at this speed shooting was poor.

'The commander-gunner had plenty of room and with ammunition in the turret bulge he could easily turn and load the 37mm gun. He could not, however, load and clear stoppages on the co-axial 7.92mm MG. His foothold on the floor was insecure and in certain positions of the turret he had to crouch owing to the height of the propeller shaft.

'A sling seat suspended by three chains (one between the gunner's legs) was provided. The adjustment, though simple, could not be altered without standing up. The seat was unsuitable for firing on the move and the gunner could not lift his body up and down to follow the relatively large arcs travelled by the telescope eye-piece and the shoulder-piece, both of which were some distance from the trunnions. The seat could not be used by the gunner when loading for himself.

'Without the seat it was easy to change from the position of commander to that of gunner and vice versa, but it must be emphasised that once the commander of a tank becomes a gunner, he ceases to command.

'The loader's position was very cramped and uncomfortable. The floor offered a very poor foothold and there were no handholds. He is liable to get part of his body caught between the 37mm gun [spent case] deflector plate and the turret roof—particularly dangerous if the elevation lock becomes jammed.

'The hull gunner's position for horizontal fire was very comfortable for a short man. By pressing with his feet on the cross-shaft casing, the gunner could brace himself against the back of the seat, which had a sliding backward and forward adjustment. The shoulder-piece on the gun was also comfortable.

'When firing at targets above or below the horizontal, however, the gunner found himself unable to follow the rather extensive movement of the telescope eye-piece and shoulder-piece. It was necessary for the gunner either to contract and extend his body like a caterpillar or, what is only slightly easier, to slide his bottom backwards and forwards by pressing with his feet or pulling on the gun.

'When travelling the gun could be released and the mounting locked but the gunner could not sit back comfortably as the wireless set was in the way of his left shoulder.

'The driver had a good seat and satisfactory controls. The position, however, was tiring for a long march as his view through the episcope was limited and he could not open any flaps. Headroom was insufficient and there was no vertical seat adjustment.'

While the author's comments had obvious validity, it must also be mentioned that in German service the commander-gunner's sling seat was replaced by a stool suspended by a bar from the turret ring, and that a firmer footing was provided for both him and the loader. The report contains numerous additional points of minor criticism, some of them trivial, and as the vehicle left Lulworth the morning after the trial shoot it can safely be assumed that the decision had already been made not to proceed with the purchase. This is hardly surprising, for March 1939 was the month that Hitler consolidated his hold over the whole of Czechoslovakia, and the Royal Armoured Corps' equipment requirements could scarcely have been left to the whims and fancies of the Führer.

A column of PaK 4.7cm(t) Sfl guns in Belgium, May 1940. Combining the PzKpfw I Ausf.B chassis with the potent Czech 47mm anti-tank gun in a partially protected mounting, they later served in both Russia and North Africa. (US Nat. Archives)

PzKpfw 38(t) Derivatives

The standard PzKpfw 38(t) could quickly be converted to the command role by installing extra radios and fitting a frame aerial above the engine deck, and in this version the vehicle was known as **Panzerbefehlswagen 38(t)**.

Once the T-34 had been encountered on the Eastern Front it was obvious that the PzKpfw 38(t)'s days as a gun tank were numbered, although many continued to serve with Germany's satellite armies well into 1942. Obsolete though the vehicle might be, its robust chassis was capable of conversion to a variety of roles and remained in production until 1944. Perhaps the simplest conversion of all was the **Aufklärungspanzer 38(t)**, in which the original turret was replaced by that of a SdKfz 222 armoured car. Seventy of these vehicles were built in 1944 and issued to favoured armoured reconnaissance battalions (see also Vanguard No 25, *German Armoured Cars & Reconnaissance Half-Tracks*).

The most common conversions, however, were to the role of self-propelled gun carriage. The **Marder III** tank destroyer appeared in three forms, one of which mounted the Russian Model 36 76.2mm anti-tank gun, while the other two carried the German 75mm L/46 PaK 40/3 anti-tank gun, one with a conventional rear-mounted engine and the other with the engine moved forward. A more advanced, fully armoured tank destroyer design was the low-slung **Hetzer**, which was equipped with a 75mm L/48 PaK 39 anti-tank gun. Fuller details of these tank destroyers and their active service can be found in Vanguard No 12, *Sturmartillerie and Panzerjäger*. A small number of Hetzers were built with a disguised flame gun in place of the 75mm main armament, their official title being **Flammpanzer 38(t)**. The flame gun had a range of 66 yards and 154 gallons of fuel were carried for it. These vehicles were used in the 1944 Ardennes offensive.

The PzKpfw 38(t) chassis also provided two carriages for the 150mm s.I.G. 33/1. Both were designated SdKfz 138/1 and sometimes referred to as the **Bison**, but while the Ausf.H version retained the central fighting compartment and rear engine, the Ausf.M saw the engine moved forward and the fighting compartment re-located at the back of the vehicle. Both models served in the Heavy Gun Companies of Panzer Grenadier divisions, together with an ammunition carrier based on the Ausf.M.

The **Flakpanzer 38(t)** was a stop-gap anti-aircraft vehicle of which 162 were built in 1943. Like the Bison Ausf.M its engine was mounted forward and the fighting compartment overhung the rear of the chassis. Armament consisted of a single 20mm cannon for which 540 rounds of HE tracer and AP tracer ammunition were stowed. The gun had a maximum rate of fire of 480 rounds per minute, but was usually governed to half this output. The vehicle was manned by a crew of five, and served in the AA platoons of armoured regiments until replaced by Flakpanzer IV models (see Vanguard No 18, *The Panzerkampfwagen IV*).

Altogether, the PzKpfw 38(t) was used as the basis for nearly 3,700 self-propelled carriages and 102 ammunition carriers. With the exception of the Hetzer series, fighting compartments were all open-topped structures of thin armour plating. In addition, some chassis were used as mobile smoke-screen layers (**PzKpfw 38(t) mit Nebel Ausrastung**), while others were sent to tank driving schools where they were converted to burning wood gas generated in a bulky apparatus installed at the rear of the vehicle. From August

The PzKpfw I Ausf.A (LaS) was an open-topped chassis widely used by the Army and NSKK for primary driver training. These vehicles were photographed as late as 1943; note M43 cap worn by officer or official at left, 1943 dull yellow factory paint scheme, and gas generators mounted on rear decks. (US Nat. Archives)

1944 a small number of unarmed Hetzers were converted to the Bergepanzer (armoured recovery vehicle) role, a 2-ton derrick being mounted on the vehicle's roof.

The PzKpfw 38(t) chassis was also widely used on experimental projects, very few of which left the drawing board. Some would have employed a German re-design which was driven by a more powerful 210hp Tatra diesel engine located beside the driver. This development was known as the PzKpfw 38(d)—d = 'Deutsch'—but the war ended before it entered production.

It speaks volumes for the chassis' reputation that it was still actively employed until the 1970s, being used by the Swedish Army for its APC **Pansarbandvagn 301**, while an improved version of the Hetzer, the **Panzerjäger G 13**, served with the Swiss Army for many years after the end of World War II.

Organisation

The concept of the Panzer division was that of an all-arms battlegroup in which those arms supporting the tank element were given a comparable mobility. As initially planned, the division's cutting edge was a Panzer brigade two regiments strong; each regiment consisted of two battalions subdivided into four fighting companies each, giving the brigade a total theoretical tank strength of 562. Three of these companies were equipped with a mixture of PzKpfws I, II and III while the 4th or Heavy Company was equipped with the PzKpfw IV.

The 1st, 2nd and 3rd Panzer Divisions were formed in October 1935, while the 4th and 5th followed in 1938. As far as the armoured brigades were concerned, they are best regarded as reservoirs which slowly filled with vehicles coming on stream from the factories and with personnel who had completed their technical training, but which never achieved their full capacity; indeed, on the outbreak of war Panzer battalions were forced to leave the officers and men of their 3rd Companies behind in their depots, so acute was the vehicle shortage. Thus, while the brilliantly staged Nazi Party Rallies and Hitler Birthday

The VK 1801 was the prototype for the 'PzKpfw 1 nA verst', the up-armoured infantry support derivative; note wider tracks and interleaved suspension. (Imp. War Mus.)

Parades, with their serried ranks of PzKpfw Is and IIs advancing towards the rostrum and parting before it, created the intended impression of invincible armoured might, during the early years of the Panzerwaffe they represented not the tip of the iceberg but the iceberg itself.

In addition, the Panzerwaffe's wish to control its own destiny had not gone unchallenged by the influential infantry and cavalry lobbies. The infantry wanted tank support for their operations, and two independent Panzer brigades (4th and 6th) and one independent Panzer regiment were formed for this purpose. The cavalry, most of which had been mechanised, was still employed on its traditional tasks of strategic reconnaissance and flank protection, and also wanted tanks for its four Light Divisions, which were formed in 1938. These divisions included one Panzer battalion, four motor rifle battalions (known as Kavellerie Schutzen to commemorate their traditions) and reconnaissance, artillery and engineer elements. The problem was that infantry and cavalry requirements absorbed tank production which would otherwise have benefited the Panzer divisions.

However, in 1938 Guderian was appointed Chief of the Inspectorate of Mobile Troops, and since the Light Divisions now came within his sphere of authority he was able to control the direction which the Panzerwaffe was to take. His intention was to increase the strength of the Light Divisions until they attained full Panzer status, but this ambition had only partially been realised by the time the Polish campaign began. Secondly, such commitments as the Panzerwaffe had to

Theoretical Organization of Panzer Division, 1939

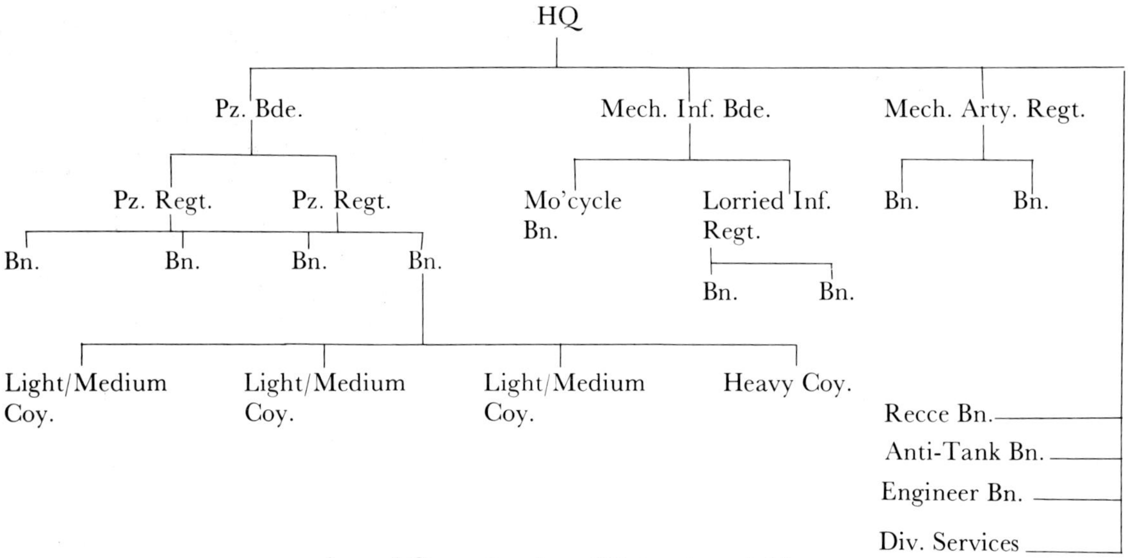

```
HQ
├── Pz. Bde.
│   ├── Pz. Regt.
│   │   ├── Bn.
│   │   │   └── Light/Medium Coy.
│   │   └── Bn.
│   │       └── Light/Medium Coy.
│   └── Pz. Regt.
│       ├── Bn.
│       └── Bn.
│           ├── Light/Medium Coy.
│           └── Heavy Coy.
├── Mech. Inf. Bde.
│   ├── Mo'cycle Bn.
│   └── Lorried Inf. Regt.
│       ├── Bn.
│       └── Bn.
├── Mech. Arty. Regt.
│   ├── Bn.
│   └── Bn.
├── Recce Bn.
├── Anti-Tank Bn.
├── Engineer Bn.
└── Div. Services
```

Actual Organization of Brigades within Panzer Divisions, 1939/40

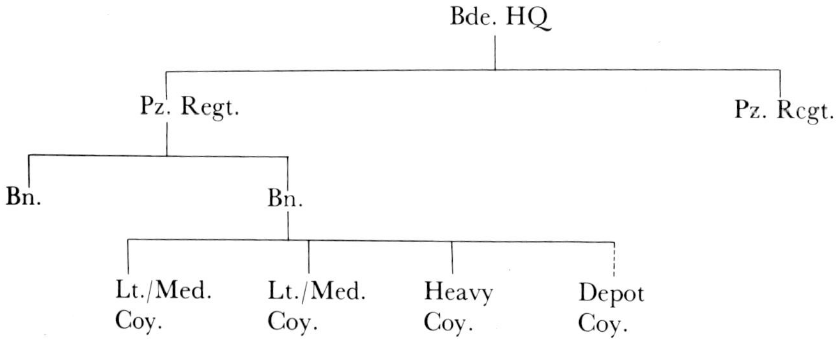

```
Bde. HQ
├── Pz. Regt.
│   ├── Bn.
│   └── Bn.
│       ├── Lt./Med. Coy.
│       ├── Lt./Med. Coy.
│       ├── Heavy Coy.
│       └── Depot Coy.
└── Pz. Regt.
```

Organization of Light/Medium Panzer Company, 1940/41*

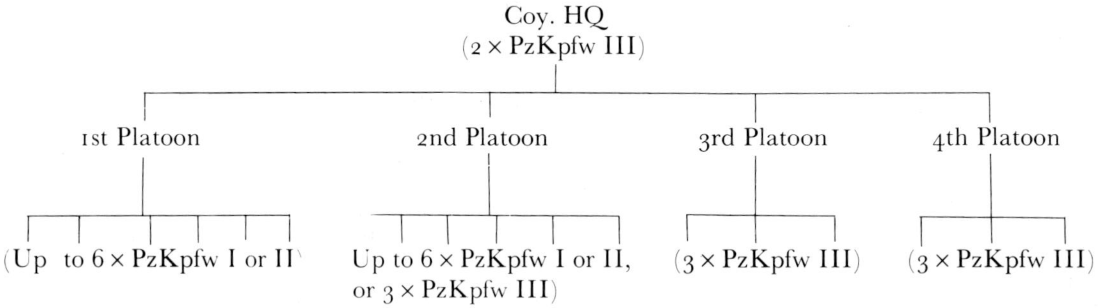

```
Coy. HQ
(2 × PzKpfw III)
├── 1st Platoon — (Up to 6 × PzKpfw I or II)
├── 2nd Platoon — Up to 6 × PzKpfw I or II, or 3 × PzKpfw III
├── 3rd Platoon — (3 × PzKpfw III)
└── 4th Platoon — (3 × PzKpfw III)
```

*In practice organization varied widely, depending upon available equipment.

Two views of the 'PzKpfw I nA verst' in service. (Above) **is an interesting study of a vehicle running on a narrower track than was usually employed for this 18-ton tank, and consequently the two overlapping bogies have been removed note protruding stub axles. Beside the driver's visor is a battery of three grenade dischargers of the type later fitted to the early Tigers.** (Right) **is a side view of the tank captured on the Russian Front while serving with 1.Pz-Div., 'red 25'—see Plate E1.** (Imp. War Mus. and J. Grandsen)

infantry support were firmly pushed in the direction of the embryo Sturmartillerie, soon to become an élite branch of service within its own right. The independent Panzer brigades were broken up and their regiments sent to reinforce the Light Divisions; one, Panzer-Regiment 8, provided the nucleus of 10.Panzer-Division, which was formed in April 1939. When Germany invaded Poland in September 1939 the Panzerwaffe could field six Panzer and four Light Divisions, plus an independent battlegroup based on I/Panzer-Regiment 10 and known as Panzerverband Ostpreussen.

Following the Polish campaign the Light Divisions were re-designated Panzer Divisions and numbered as follows:
1.Leichte-Division became 6.Panzer-Division, with Panzer-Regiment 11 (two battalions) and Panzer-Abteilung 65.
2.Leichte-Division became 7.Panzer-Division, with Panzer-Regiment 25 (two battalions) and Panzer-Abteilung 66.
3.Leichte-Division became 8.Panzer-Division, with Panzer-Regiment 10 (two battalions) and Panzer-Abteilung 67.
4.Leichte-Division became 9.Panzer-Division, with Panzer-Regiment 33 (two battalions).

All ten Panzer divisions took part in the 1940 campaign in the West, their performance impressing Hitler to such an extent that in August of that year he decided to raise eleven more, numbered 11 to 20 inclusive and 23. This was achieved by reducing the tank element of each division to a single regiment, but since existing divisions were already below their established tank strength this further dilution is best regarded as an act of folly which struck at the Panzerwaffe's principal tenet of concentration. The decision also placed an impossible strain on the overtaxed German armaments industry; to illustrate the point, 23. Panzer-Division was not adequately equipped until October 1941.

The rout of Hitler's Italian allies in North Africa led to the hasty formation of the 5.Leichte-Division in February 1941, based on a cadre from 3.Panzer-Division. In July 1941 5th Light changed its title to 21.Panzer-Division. The 90th Light (Afrika) Division was a motorised infantry formation which in March 1942 became simply the 90th Infantry Division (Motorised).

This continued expansion alone prolonged the active service life of the German and Czech light tanks, which were still in a majority during the first months of Operation 'Barbarossa'. Thereafter, those that survived were withdrawn from the Panzer battalions for conversion to other uses. Tremendous efforts made by the armaments industry to fill the gap were thwarted by Hitler's decision to raise yet more Panzer divisions, not only for the Army but also for the Waffen-SS.

Active Service

Hitler's resolution to support Franco during the Spanish Civil War led to the despatch of the 'volunteer' Legion Condor, incorporating Air Force and Army elements, and the Panzerwaffe received its baptism of fire in January 1937. The Legion's tank unit was armed with 100 PzKpfw Is

PzKpfw II Ausf.B of Pz-Regt.36, 4.Pz-Div. in France, 1940. The Ausf.C was the last model to be produced with the original thinly-armoured, rounded bow. Experience in Poland showed it to be vulnerable even to anti-tank rifle fire, and eventually many of the early models were retrospectively fitted with a squared-off appliqué armour kit similar in appearance to the later variants. (US Nat. Archives)

and commanded by the then Major Wilhelm Ritter von Thoma.[1] In Spain the unit was able to develop operational techniques in an active service environment, particularly co-operation with the ground support wings of the Luftwaffe, and the forward deployment of an anti-tank gun screen through which the tanks could retire if hard pressed by the much superior Russian T-26s. The concepts of mobility and concentration were also vindicated; but Spain was not a representative proving ground for the protagonists of the Blitzkrieg theory and no dramatic successes were recorded. Indeed, von Thoma's reports suggested that the tank had failed to live up to its promise; but Guderian, recognising the fact that this was not a conventional war fought between first rate national armies, refused to be deflected from his beliefs.

Those beliefs were to be tested again during the Austrian Anschluss of March 1938, when up to 30 per cent of the tanks involved in the march on Vienna littered the roadside, broken down, although the major part of Guderian's 2.Panzer-Division succeeded in covering 420 miles in only two days. Once installed in his Inspectorate, Guderian implacably set about eradicating failures in the tank recovery and repair system. The subsequent occupation of the Sudetenland (October 1938) and Czechoslovakia revealed a greatly improved situation, and in this context it is worth remembering that the PzKpfw I and II were light enough to be ferried long distances on unadapted lorries into their operational zones, so saving critical track mileage.

Whether Hitler would have continued to play his deadly game of brinkmanship over the Polish question had he not acquired the Czech armoury remains a matter of speculation. In the event his bluff was called, and to its horror the Wehrmacht found itself facing the possibility of war on two fronts. Not even the most enthusiastic officers of the Panzerwaffe could have welcomed the prospect of going to war with their under-strength, badly equipped divisions.

---

[1] General der Panzertruppe von Thoma was captured during the closing hours of Second Alamein.

The nominal tank strength of a Panzer division in 1939 was 562 vehicles: the reality was somewhat different. As the senior formation, 1.Panzer-Division received a larger allocation of the few medium tanks available, each of its battalions containing 14 PzKpfw IVs, 28 PzKpfw IIIs, 18 PzKpfw IIs and 17 PzKpfw Is; together with command and headquarters vehicles the divisional tank strength was 324. The battalion establishments of the remaining Panzer divisions was six PzKpfw IVs, five PzKpfw IIIs, 33 PzKpfw IIs and 34 PzKpfw Is; these, together with command vehicles, produced a tank strength of 328 per division. The 1.Leichte-Division, with three tank battalions, could muster a total of 221 tanks, including a handful of PzKpfw IIIs and IVs, but was one of the strongest formations in the field since the balance of its tank strength included 112 PzKpfw 35(t)s. Meanwhile 2. and 4.Leichte-Divisionen each had less than 100 tanks, and of these the PzKpfw II predominated. The 3.Leichte-Division, however, was reinforced with a battalion of 59 PzKpfw 38(t)s, giving it approximately 150 tanks.

It was, therefore, upon the light tanks that the burden of the Polish campaign fell. In this context it must be remembered that in 1939 few soldiers had experienced an armoured attack, and that to the average infantryman a tank was a tank, extremely dangerous and ostensibly invulnerable, whatever shortcomings its users may have felt it had. Again, the Polish Army was organised and trained for operations in the style of 1918, and its own armoured force was small and much of its equipment obsolete; Polish sources are scrupulously honest in stating that insofar as their armour was concerned the battle could never have been won, but it could have been fought better.

The course of the campaign will be known to most. Poland, strategically outflanked by East Prussia in the north and Czechoslovakia in the south, was subjected to concentric attacks in the form of two giant pincer movements, the inner set to close near Warsaw and the outer near Brest Litovsk. Once the cordon of Polish armies guarding the frontier had been penetrated the continued advance of the Panzer Korps totally disrupted their command and logistic networks, inducing eventual collapse. The *coup de grâce* was treacherously administered by the Red Army, the Soviet government claiming that it had 'intervened' to stop the fighting, but keeping a huge slice of Polish territory as a reward for its benevolence.

The Panzerwaffe emerged from the campaign with its confidence enhanced and its operational techniques sharpened by experience; but it had by no means enjoyed the 'dry run with a little shooting' that was later suggested. In some areas senior commanders had been forced to apply considerable pressure to overcome their formations' natural reluctance to maintain their drive into unknown territory in accordance with the theory of deep penetration. The mechanical failure rate, particularly among the PzKpfw Is and IIs, rapidly rose to 25 per cent and remained there. Above all, whenever possible the Poles had fought back with a desperate bravery, using their few anti-tank guns, anti-tank rifles and field

PzKpfw II Ausf.D tanks using mobile ramps to clamber aboard their transport vehicles. One of the better features of German and Czech light tank designs was their easy portability on commercial lorries, obviating the need for expensive specialised transporters to carry them long distances outside the combat zone. (RAC Tank Museum)

artillery firing over open sights; 4.Panzer-Division alone lost 60 tanks in a single day, trying to fight its way prematurely into central Warsaw. For the campaign as a whole, the Panzerwaffe admitted a loss of 218 tanks, approximately 10 per cent of the total engaged: the breakdown of total losses was 89 PzKpfw Is, 78 PzKpfw IIs, 26 PzKpfw IIIs, 19 PzKpfw IVs, and six PzKpfw 35(t)s. In the light of subsequent strength returns this figure has been regarded as suspect, and a post-war Polish examination of contemporary German documents reveals it to be so. The researchers' findings were published in the reputable *Wojskowy Przeglad Historyczne* (Military History Journal) and I am grateful to Steven Zaloga for drawing my attention to the relevant article. What actually appears is a reduction in the Panzerwaffe's operational tank strength by 674 vehicles, partly the result of battle damage, and partly because of mechanical failure and other causes. If one accepts that one third were immediately written off, then one arrives close to the published German figure. A further third should simply be regarded as being beyond local repair, and these would have been re-built in Germany. The remainder, whether damaged in battle or suffering from major mechanical defects, fall into the category of being beyond economic repair, and in this respect it is worth remembering that the PzKpfw Is and IIs were comparatively frail machines and that the more elderly of them had already reached the limit of their service life. The full extent of the German tank loss, therefore, almost certainly exceeded 400 vehicles, or 20 per cent of the total; and this, balanced against deliveries of new and repaired vehicles, is reflected in the line-up for the offensive in the West the following year. This might be regarded as unduly high for a month's fighting in which no major armoured engagement took place, but the fact remains that in that time the Panzer divisions and the Luftwaffe together defeated an army of 1,500,000 at a cost of only 8,000 German lives.

Hitler's next campaign of expansion was

Infantry machine-gun squad taking cover behind a PzKpfw II Ausf.B of 2.Kompanie, Pz-Abt.zBV 40 during the invasion of Norway. The battalion insignia, a yellow 'V' in a circle, can be seen left of the white outline rear hull cross. Note white rectangular air recognition panel painted on the engine deck. (US Nat. Archives)

1: PzKpfw I Ausf.A, 5 Komp., Pz-Regt.1, c.1935

2: PzKpfw I Ausf.B, 2° Bn., Agrupacion de Carros; Spain, 1938

1: KlBefwg I, 4.Panzer-Division; Poland, 1939

2: PzKpfw I Ausf.B, 4.Panzer-Division; France, 1940

1: PzKpfw II Ausf.b, Pz-Regt.7, 10.Pz-Div.; France, 1940

2: PzKpfw II Ausf.C, 4.Panzer-Division; France, 1940

1: PzKpfw II Ausf.B, Pz-Aufkl-Abt.2, 12.Pz-Div.; Russia, 1942

2: PzKpfw II Ausf.F, Slovak 1st Inf.Div.; Russia, 1943

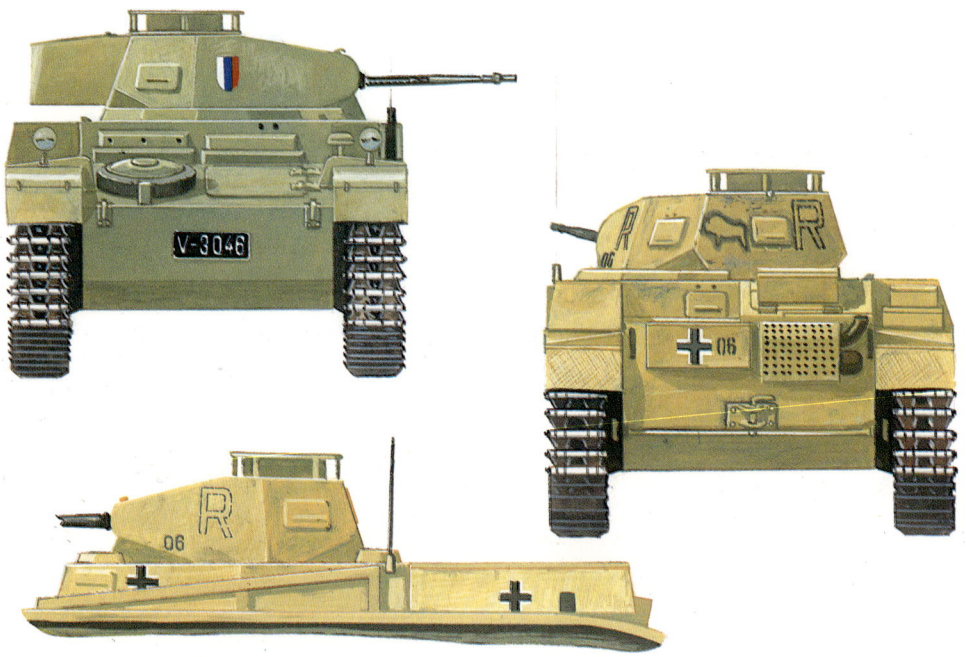

3: PzKpfw II Ausf.F, HQ Pz-Regt.7, 10.Pz-Div.; Tunisia, 1943

D

1: PzKpfw I Ausf.F, 1.Pz-Div.; Russia, 1943

2: PzKpfw II Ausf.L 'Luchs', unit unknown; Normandy, 1944

1: PzKpfw 35(t), II/Pz-Regt.11, 6.Pz-Div.; Russia, 1941

2: LTvz35, Rumanian 1st Tank Regt.; Stalingrad front, Nov. 1942

1: PzBefw 38(t), 7.Panzer-Division; France, 1940

2-10: Flags and pennants – see Plates commentary for details.

1: PzBefw 38(t) Ausf.B command tank, 8.Pz-Div.; Russia, 1941

2: LTvz38, 11th Tank Co., Slovak Mobile Div.; Russia, 1941

PzKpfw II of Panzergruppe Kleist—note white 'K' on rear hull—crossing a stream in France, 1940. The detachable plate bearing the tank's tactical number is retained here. The crew have added a wooden 'fence' to the left of the engine deck to keep exterior stowage in place. The combination of field grey sidecaps and black uniforms was characteristic of the French campaign. (US Nat. Archives)

directed at Norway and Denmark under the codename of '*Weserübung*' (Exercise 'Weser'). This began on 9 April 1940, and although the initial assault was delivered by airborne and air-landing troops, a tank battalion known as Panzer-Abteilung z.B.V. 40 was included in the second echelon. This battalion, formed from Panzer-Regiment 35 of 4.Panzer-Division, reflected in its equipment the chronic tank shortage still afflicting the Panzerwaffe, now preparing for what was regarded as its supreme test, the invasion of western Europe. None of the scarce PzKpfw IIIs and IVs could be spared for '*Weserübung*', but in their place were respectively 15 PzKpfw 38(t)s and three Neubaufahrzeug prototypes which, like the PzKpfw IV, carried a 75mm main armament; the remainder of the battalion consisted of PzKpfw Is and IIs, approximately 40 in number.[1]

The Danes could offer only a token resistance to the onslaught; the main body of the battalion advanced through Denmark and was then shipped over to Oslo, where it joined the NbFz platoon, which had landed as soon as the city had been secured. The tanks were then employed in spearheading the German drive up the two principal valleys north of Oslo, Gudbrandsdalen and Osterdalen. In Gudbrandsdalen the Panzers' advance was delayed by demolitions and brought to a halt by the stiff resistance offered by the British 15th Infantry Brigade. Two vehicles were destroyed by the brigade's French 25mm anti-tank guns, but the Luftwaffe's intervention and the growing volume of German artillery fire at length caused the British to withdraw. In Osterdalen the tanks faced only lightly equipped Norwegian troops and sustained no major checks. By the beginning of May southern and central Norway was firmly in German hands.

On 10 May 1940 the long-awaited offensive

[1] The Neubaufahrzeug (NbFz) was an abandoned design for a multi-turreted breakthrough tank.

against France and the Low Countries began. An appendix to Guderian's memoirs contains a memorandum despatched by Army Headquarters (OKH) to the Führer's military adjutant on 7 November 1944, and this quotes the German tank strength on the eve of the invasion as being:

PzKpfw I	... 523	PzKpfw 35(t)	... 106
PzKpfw II	... 955	PzKpfw 38(t)	... 228
PzKpfw III	... 349	PzBefw I	... 96
PzKpfw IV	... 278	PzBefw III	... 39

These figures, prepared four and a half years after the event, were as accurate as possible, although there is good reason to believe that they slightly understate the case insofar as the Czech vehicles are concerned. Excluding command vehicles, they give a total of 2,439 gun tanks, of which the 1,478 machine-gun-armed PzKpfw Is and IIs still represented a substantial majority. In contrast, the French Army alone could deploy over 3,000 tanks, many of which were better armed and better protected than the best German machines.

The strategic plan for the campaign was the brainchild of the brilliant von Manstein, who predicted, quite correctly, that a German invasion of Holland and Belgium would draw the best French armies and the British Expeditionary Force north into a trap. Once the Allies had taken the bait a concentrated drive by the mass of the Panzerwaffe from the Meuse to the Somme Estuary would isolate the British, French and Belgian armies in the north and ensure their destruction. The course of the campaign itself entirely fulfilled Manstein's concept, and has been described in Vanguard Nos 16 and 18, but it is worth mentioning here the part played by the light tanks in this most sweeping of German victories.

Responsibility for the invasion of Holland and Belgium rested with von Bock's Army Group B which was allotted only three Panzer divisions—3., 4., and 9.—although to compensate for this it received most of the airborne and air-landing formations. In this context it should be remembered that any failure on von Bock's part to convince the Allies that his invasion was the main German thrust line would be reflected adversely against von Rundstedt's Army Group A in its drive to the coast.

One armoured formation upon which much

Column of PzKpfw II tanks of 7.Pz-Div.—see yellow 'Y-rune' on right corner of hull front—advance into Russia during Operation 'Barbarossa', summer 1941. These are Ausf.A-C variants displaying the new angled nose armour and the new vision cupola for the commander. (J. Grandsen)

depended was Major General Ritter von Hubicki's 9.Panzer-Division, which had been recruited in Austria. Ironically, this was the weakest of all the Panzer divisions, its Panzer-Regiment 33 having only two battalions equipped with a total of 18 PzKpfw IVs, 36 PzKpfw IIIs, 75 PzKpfw IIs and 100 PzKpfw Is; yet it was expected to drive straight across Holland and relieve the German parachute troops holding the Moerdijk bridge over the Maas before the French 1st Army, approaching from the south, could get there. In the event Hubicki's task was made easier by the French commander who, within striking distance of the objective, split his advance guard into a reconnaissance screen which was easily brushed aside, and then withdrew. 9.Panzer-Division broke through to the Fallschirmjäger on 12 May, and then swung north to Rotterdam, which was entered two days later. Although eclipsed by events elsewhere, the division's achievement was astonishing; it had cut Holland in two and isolated the Dutch Army from Allied aid and reinforcement—the entire Manstein plan in microcosm, in fact. Holland surrendered on 15 May.

Further south but still under the command of Army Group B was Hoepner's XVI Panzer-Korps, consisting of 3. and 4.Panzer-Divisionen, each of which had a nominal tank strength of 24 PzKpfw IVs, 50 PzKpfw IIIs, 110 PzKpfw IIs and 140 PzKpfw Is. After crossing the Maastricht 'Appendix' Hoepner's divisions motored on across Belgium until they reached the Gembloux Gap, and here they were met by General Prioux's 1st French Cavalry Corps, covering the flank of 1st Army as it deployed. Prioux's corps consisted of the 2e and 3e Divisions Légères Mecaniques, and was equipped with 174 of the formidable Somuas, 87 Hotchkiss H35s and 40 AMR light tanks, together with a small Belgian contingent. A fierce tank battle was fought throughout the 12–13 May, with each side suffering losses in excess of 100 vehicles and both claiming a victory. However, while the French retired within 1st Army's prepared positions, the Germans retained possession of the field and were able to recover many of their casualties. The principal effects of this engagement, referred to in German records as the 'Panzerschlacht bei Namur', were to rivet French attention on Belgium and to divert the main Allied counter-attack force, the three French Divisions Cuirassées, away from the real threat that was unfolding in the Ardennes, with disastrous consequences.

The most northerly of Army Group A's armoured formations was Hoth's XV Panzer-Korps, composed of 5. and 7.Panzer-Divisionen, the latter commanded by Major General Erwin Rommel. The former was very similarly equipped to the 3rd and 4th Divisions, but the 7th's equipment at the outset consisted of 36 PzKpfw IVs, 106 PzKpfw 38(t)s, 40 PzKpfw IIs and 10 PzKpfw Is.[1] However, while Rommel's division was moving up to the Meuse, 5.Panzer was delayed by congested roads and Hoth placed its leading armoured regiment, Panzer-Regiment 31, under his command, thus making 7.Panzer an extremely powerful division. The attachment was intended to be a temporary one, but Rommel was a thoroughly bad neighbour and it is debatable just when or how much of Panzer-Regiment 31 he returned to its owners; certainly, as late as 21 May he sustained the loss of six PzKpfw IIIs in action and these vehicles, as we have seen, were not included in his own order of battle.

In the centre of Army Group A's sector was Rheinhardt's XLI Panzer-Korps, composed of 6. and 8.Panzer-Divisionen. Both had a similar proportion of PzKpfw Is, IIs and IVs to Rommel's division, but the bulk of 6.Panzer-Division's tank strength consisted of 118 PzKpfw 35(t)s accompanied by 10 PzBefw 35(t)s, while 8.Panzer-Division had 116 PzKpfw 38(t)s and seven PzBefw 38(t)s.

On the Army Group's left was the strongest Panzer-Korps of all, Guderian's XIXth, consisting of 1., 2. and 10.Panzer-Divisionen. Guderian was to lead the drive to the sea and because of this his divisions received a generous allocation of medium tanks, their respective strength being 56 PzKpfw IVs, 90 PzKpfw IIIs, 100 PzKpfw IIs and 30 PzKpfw Is each.

These dispositions, both quantitative and qualitative, reflect Manstein's overall strategy in that the best equipment, including the reliable Czech vehicles, was concentrated on the German left. In contrast, the French disposed their armour

---

[1] Figures for Czech vehicles are as quoted in *Czechoslovak Armoured Fighting Vehicles 1918–1945* by H. C. Doyel and C. K. Kliment published by Bellona.

evenly along the front so that the Panzer-Korps, present in overwhelming strength, found no difficulty in breaking through and out into open country. In both the dash to the sea, completed on 20 May, and the subsequent operations against the remaining French armies south of the Somme, the Panzer divisions acted with a drive and confidence that had been absent in Poland. When the campaign ended on 22 June the world's two most prestigious armies had been defeated. German casualties in six weeks' fighting amounted to only 150,000; but losses in action and through sheer mechanical attrition had reduced the Panzerwaffe's tank strength by 50 per cent.

While no Czech tanks were sent to North Africa with Rommel, his command did contain a substantial number of PzKpfw Is and IIs. The former soon disappeared, but on the eve of Operation 'Crusader' (November 1941) PzKpfw IIs still constituted almost one third of the total German tank strength of 249. Thereafter, those that survived were withdrawn for conversion.

In April 1941, with planning for the onslaught on Soviet Russia well under way, Hitler launched his invasions of Yugoslavia and Greece. Neither country possessed adequate modern anti-tank equipment, but both believed that their mountainous terrain would inhibit the sort of run the Panzers had had in Poland and France. This belief proved to be mistaken, for although the terrain did present problems these were far from insuperable. Like the Polish and French General staffs, their Greek and Yugoslav counterparts were simply not prepared for the speed with which events developed once the Panzer divisions had broken through or by-passed their static defences. Yugoslavia surrendered on 17 April and Greece on the 23rd.

About one third of the German armoured force had been engaged.[1] Many senior officers believed that the Balkan campaign had been unnecessary and that the time spent in its preparation, execution

PzKpfw II marking variations. (Left) **is an Ausf.C of the signals platoon ('N' for Nachrichtung) of a second battalion headquarters ('II') of 2.Pz-Div. (two yellow dots just right of the individual tank number '3') near Sedan in 1940.** (Below left) **served with the third battalion HQ recce platoon, i.e. Stab, Pz-Abt.66, in 7.Pz-Div. in Russia, 1941. Both the black and white cross and the yellow rune divisional sign are carried on hull sides and turret rear; the 'III 14' is red outlined in white.** (Below) **served with first battalion HQ ('I') and tank number '06' in yellow) of 4.Pz-Div., whose yellow sign is marked on the hull side. The name 'Franz Lott' on the turret commemorates a dead comrade.** (US Nat. Archives)

[1] 2., 5., 8., 9., 11. and 14.Panzer-Divisionen.

Fully up-armoured Ausf.A, B or C photographed in North Africa, with insignia of the Deutsches Afrika Korps and 21. Pz-Div. clearly visible. Note large bolt-heads, confirming the application of extra plate to the hull and turret front; the reinforced mantlet with 'splash' plates; and the angular bow. (RAC Tank Museum)

and the subsequent redeployment contributed to the failure of Operation 'Barbarossa'; as indeed did the imposition of substantial track mileage, particularly on older vehicles, when the Panzerwaffe was on the point of being asked to perform the impossible.

Manufacture of German medium tanks had been accelerated after the fall of France, but the war against Russia opened with the Panzerwaffe *still* having a preponderance of light tanks, its grand total of 3,332 vehicles being made up of 410 PzKpfw Is, 746 PzKpfw IIs, 965 PzKpfw IIIs, 439 PzKpfw IVs, 149 PzKpfw 35(t)s and 623 PzKpfw 38(t)s, the Czech tanks forming the principal armament of 6., 7., 8., 12., 19. and 20. Panzer Divisionen.[1] Altogether, 17 Panzer divisions, with two more in reserve, spearheaded the invasion, organised into four Panzergruppen.

[1] 6.Panzer-Division had 149 PzKpfw 35(t)s; the remainder PzKpfw 38(t)s, viz: 7.-167; 8.-118; 12.-107; 19.-118; 20.-113.

The sheer arrogance of Hitler's ambition was terrifying. The Red Army possessed over 20,000 tanks, unlimited manpower resources and an infinity of space into which it could withdraw. Unlike France, where the strategic objectives were just mechanically attainable but which placed serious strain on the repair organisation, those in Russia were many times further away; moreover, surfaced roads were the exception in Russia, making the going even more difficult.

Yet the Panzergruppen, now thoroughly experienced and at the peak of self-confidence, performed prodigies. Russian armies were repeatedly encircled and destroyed and Moscow was within their grasp when, at the critical moment, Hitler intervened needlessly to divert the central thrust to entrap the already crumbling Soviet formations retreating before Army Group South. By the time the advance was resumed the Panzer spearheads had been eroded to mere shadows of their former selves, while Stalin was rushing Siberian reinforcements to the Moscow front. Worse, the autumn rains created a sea of mud in which only tanks could move and then only with difficulty; many of the less combat-worthy PzKpfw Is and IIs found themselves towing trains of three or four lorries through axle-deep slime so that communications could be kept open.

With the first frosts the ground hardened, and the advance continued for a while. It was finally halted on 8 December by deep snow and unimaginable cold, crews and their machines having to cope with conditions which exceeded the wildest visions of a Nordic hell. In the low temperatures track pins broke, firing pins shattered, telescopic sights iced over, recoil buffer fluid became solid and so prevented the firing of guns, and sump oil assumed the viscosity of treacle. Fires had to be kept burning beneath tanks, and engines had to be run every four hours throughout the night.

True, the Red Army had sustained the staggering loss of 17,000 tanks and over one million men; but German casualties (excluding the numberless frostbite cases) amounted to 743,000, 25 per cent of those engaged. The Panzerwaffe had lost 2,700 vehicles, the repair and replacement organisations' inability to cope with the scale of the problem being compounded by Hitler's insistence, even in this moment of supreme crisis, on retaining new tanks in Germany with which to form yet more Panzer divisions.

'*Barbarossa*' had been the light tanks' swan song. Some of the early encounters with the Russian T-34s and KVs are detailed in Vanguard No 14 *The T-34 Tank* and Vanguard No 24 *Soviet Heavy Tanks*; these led at least one Panzergruppe commander to the conclusion that in future no tank with less than a 75mm main armament could hope to survive on the Eastern Front. Symbolically, the departure of the light tanks was accompanied by that of Guderian, dismissed following a violent personality clash with his immediate superior, Field Marshal Gunther von Kluge. The Years of Victory were over.

Specifications

Panzerkampfwagen I Ausf.B
Length: 4.42m Width: 2.06m Height: 1.72m Weight: 5.8 tons Armour: 13mm Armament: 2 × 7.92mm MG Crew: 2 Speed: 40kph

Panzerkampfwagen II Ausf.F
Length: 4.81m Width: 2.24m Height: 1.98m Weight: 9.35 tons Armour: 30mm Armament: 1 × 20mm cannon, 1 × 7.92mm MG Crew: 3 Speed: 40kph

'Lynx' armoured reconnaissance vehicle, apparently of the same unit as that in Plate E2—note turret number '4121'. Captured in Normandy, it is finished in dull yellow with brown and green random overspray. (RAC Tank Museum)

Panzerkampfwagen II Ausf.D
Length: 4.64m Width: 2.24m Height: 2.02m
Weight: 10 tons Armour: 30mm Armament: 1 × 20mm cannon, 1 × 7.92mm MG Crew: 3 Speed: 55kph

Panzerkampfwagen 35(t)
Length: 4.45m Width: 2.14m Height: 2.20m
Weight: 10.5 tons Armour: 25mm Armament: 1 × 37mm gun, 2 × 7.92mm MG Crew: 4 Speed: 35kph

Panzerkampfwagen 38(t)
Length: 4.90m Width: 2.06m Height: 2.37m
Weight: 9.725 tons Armour: 25mm Armament: 1 × 37mm gun, 2 × 7.92mm MG Crew: 4 Speed: 42kph

The Flammpanzer II 'Flamingo', with some detail visible of the large side fuel containers, two small remotely-controlled flame-gun turrets at the nose, and small main machine gun turret. The short range of the flame projectors required the fitting of grenade dischargers on the trackguards, just visible here behind the fuel containers, to cover the advance on the target. Seen here in action on the Leningrad front in 1941, the Flamingo was not particularly successful, and was withdrawn in March 1942. (J. Grandsen)

The Plates

A1: PzKpfw I Ausf.A, 5 Kompanie, Panzer-Regiment 1, 1935

Photographed at a Berlin parade in 1935 or 1936, this tank is finished in the 1935 regulation camouflage of Panzer grey with roughly one-third of the surface area overpainted with dull brown 'cloud' pattern. The national cross does not seem to be carried. The number '5' in yellow is carried by a whole company in the photo we copy; all foreground vehicles, in more than one platoon, also carry the black 'I' on the white parallelogram tank symbol, and this would suggest a battalion identification—but there were four companies per battalion at this date.

A2: PzKpfw I Ausf.B, 2° Batallon de la Agrupacion de Carros; Spain, 1938

The first 32 PzKpfw Ausf.A tanks were sent to Spain with von Thoma in September 1936, later being followed by Ausf.Bs, the number finally

This head-on view of the PzKpfw 35(t) in its original Czech Army livery of ochre, green and brown gives a clearer idea of the scheme seen on tanks of the Slovak Mobile Division — see Plate H2. (RAC Tank Museum)

reaching about 100 vehicles. Initially equipping the three Lehrkompanien of Panzer-Abteilung 88 of the Legion Condor, they were later used by mixed Nationalist units formed around German cadres. Initially a plain Panzer grey scheme was used, but later irregular camouflage stripes of sandy brown were added to some tanks; their nickname in Spain was *negrillos*, indicating that the grey was at first very dark, but colours faded in time. The Legion Condor units used a death's-head insignia, but with transfer to Spanish units more elaborate markings were substituted. The scarlet and yellow flashes were the Nationalist flag colours. The St. Andrew's cross was carried on the turret roof as a Nationalist air recognition symbol. The 1ᵃ Batallon de Carros de Combate del Ejercito Nacional used a halved disc in white over a second colour: red and yellow indicated 1st and 2nd Companies respectively, and all-white, 3rd Company. The 2°Bn. of the expanded Agrupacion de Carros used the same system but a diamond-shaped patch: this is therefore a 1st Company tank. The individual tank number appeared in white on glacis and hull rear, and the white trophy of arms symbol is the *Escudo de la Legion*, the badge of the Spanish Foreign Legion, which seems to have provided at least some sub-units.

B1: Kleiner Panzerbefehlswagen, 4.Panzer-Division; Poland, 1939

A command tank, carrying the usual regimental headquarters tactical numbers on the turret side, and on the number plate on the rear hull as well—'R02'. The 4.Pz-Div. was one of the few formations to display a divisional insignia in Poland, a three-point star or 'caltrop' shape painted here on the rear of the cupola. On 1 September 1939 the division lost over 40 armoured vehicles in a sharp encounter with the Polish Wolynian Cavalry Brigade near Mokra, and thereafter began either painting out or obscuring the white national cross displayed in Poland, which made too good an aiming-mark for enemy anti-tank gunners. In

PzKpfw 35(t) tanks of 6.Pz-Div. advancing as part of the northern thrust into Russia in 1941; on the original the two yellow 'Xs' of the divisional sign can be made out on the front plate left of the antenna base. The crews of this division were generally very satisfied with the performance of their 'Skodas', although the appearance of the T-34 and KV-1, together with the savage conditions of winter 1941–42, finally left them without a single tank by the early months of 1942. (US Nat. Archives)

Fine side elevation study of the PzKpfw 38(t), here a captured example to which the white serial 'EW2222' had been added. Note the commander's panoramic telescope just forward of the cupola.

this case it seems to have been smeared with mud; some seem to have been 'scribbled' over with charcoal.

B2: PzKpfw I Ausf.B, Aufklärungs-Zug, II Abteilung Stab, Panzer-Regiment 36, 4.Panzer-Division; France, 1940

The turret codes 'IIL' indicate that the 2nd Battalion was a 'light' battalion; '5' is the individual tank number. Note that these, and the 1940 version of the divisional insignia, are in yellow, while the national cross is in white outline only on the Panzer grey ground. The battalion HQ Roman 'II' was repeated on both turret sides as well as the rear.

C1: PzKpfw II Ausf.b, Panzer-Regiment 7, 10. Panzer-Division; France, 1940

The traditional regimental insignia, a bison, was marked in an unusual way by spraying round the edges of a solid stencil with white or light grey paint. Unusually, the turret tactical number is limited to the '5' indicating the company; while the full company, troop and tank number '542' is displayed on the detachable rhomboidal tin plate fixed to the hull side, and probably to the hull rear as well (see C2). The 10.Pz-Div. at this date sometimes displayed the yellow insignia shown at the far right end of the front hull plate as viewed—a Y-rune and three strokes, in yellow.

C2: PzKpfw II Ausf.C, 2.Kompanie, Panzer-Regiment 36, 4.Panzer-Division; France, 1940

The tactical number '216' on the turret is followed, in both side and rear positions, by a yellow dot; this was used within the division to indicate Pz-Regt.36, as opposed to Pz-Regt.35. National crosses appear in white, as does the tactical number on the rhomboidal plate retained at the rear; the divisional insignia on the turret rear is in yellow. The engine deck bears a broad white stripe as an air recognition sign. Note 'broken' presentation of national cross on hull side/antenna trough.

D1: PzKpfw II Ausf.B, Panzer-Aufklärungs-Abteilung 2, 12.Panzer-Division; Russia, 1942

Unusual markings are displayed by this tank of a reconnaissance battalion: a large yellow 'A (for Aufklärungs) 94' on the turret rear, repeated on the forward part of the hull side plate ahead of the 12.Pz-Div. insignia and the national cross in white outline. Heavy rear stowage includes a large crate and a bundle of logs for a fascine; and note smoke-

35

The most noted users of the PzKpfw 38(t) in 1940 were the crews of Gen. Rommel's reinforced 7.Panzer-Division. The reversed 'Y-rune' and three dots, used as the divisional sign in that campaign and visible beside the machine gun on the nearest tank here, was later changed to a simple upright 'Y' to avoid confusion with 4.Pz-Div.'s reversed 'Y' and three strokes. The turret numbers are red outlined in white, and the hull crosses black and white. (US Nat. Archives)

bomb dischargers added to each rear corner of the trackguards.

D2: PzKpfw II Ausf.F, Slovak 1st Infantry Division; Russia, 1943

This tank, photographed in the Crimea, appears to belong to 11 Tank Company shortly before that unit's disbandment—see H2. It is finished in plain dark yellow, the factory paint scheme for all German armour from February 1943 onwards. Markings are limited to the national shield on the turret sides, and a white-on-black front number plate 'V-3046'.

D3: PzKpfw II Ausf.F, Stab Panzer-Regiment 7, 10.Panzer-Division; Tunisia, 1943

This tank, now displayed in the RAC Tank Museum, Bovington, is finished in pale yellow sand camouflage all over. The 'R' of the regimental staff appears on both sides and rear of the turret; the individual tank number '06' is painted on turret sides and hull rear; and on the turret rear the regimental bison symbol is stencilled in black or dark grey outline. Note double presentation of national cross on hull sides.

E1: PzKpfw I Ausf.F, 1.Panzer-Division; Russia, 1943

The 'PzKpfw I nA verst' born of the VK 1801 prototype was built in only 30 examples and issued as the 'PzKpfw I Ausf.F'. This very heavily armoured infantry support vehicle bore little resemblance to the other PzKpfw I models. The trials unit served in Russia in 1943 with 1.Panzer-Division, where this tank, 'red 25', was captured. The tactical number was painted on the turret rear only; and the use of the 'Berlin bear' motif, normally associated with 3.Panzer-Division, is unexplained. The survivors of the trials unit were apparently transferred to anti-partisan duty in Yugoslavia, where another tank was captured by Tito's forces.

E2: PzKpfw II Ausf.L 'Luchs', unit unknown; Normandy, 1944

The heavy modification and stowage of this model left little room for displaying markings. Finished, like E1, in overall factory dark yellow with a random oversprayed mottle of dark green and red-brown, it served with an as yet unidentified reconnaissance unit in Normandy—the four-digit tactical number '4134' recalls the numbering style of armoured cars such as the Puma. In this case the number is split into two pairs, marked on a stowage box or jacking block at the rear of the turret side, and on a metal plate welded to a jerry-can rack further forward. The black and white national cross is displayed high at the centre of the rear hull plate; and another photo shows that it was painted on the central hull side stowage bin on the left side, as was normal in this unit, though this tank lacks the right hand presentation.

F1: PzKpfw 35(t), II Abteilung, Panzer-Regiment 11, 6.Panzer-Division; North Russia, autumn 1941

Details of the operations carried out by this division's Skodas will be found in Vanguard 28 *The 6th Panzer Division 1937–45*, together with monochrome and colour photographs. The usual Panzer grey finish is brightened by a large air recognition flag over the heavy stowage on the engine deck. The tactical number '713' of this tank of 7.Kompanie, 1.Zug is painted in small yellow characters centrally on the sides and rear of the turret. The national cross—sometimes in black and white, sometimes in white outline only—appeared on the hull rear and well forward on the hull side plates, though often obscured by stowage. The divisional insignia, a reversed Y-rune and two dots in yellow, appeared at the left upper corner of the rear plate, and at the right end (as viewed) of the front plate above the visor.

F2: LTvz 35, 1st Tank Regiment, Rumanian 1st Armoured Division 'Greater Rumania'; Stalingrad front, November 1942

One of the formations of Dumitrescu's 3rd Army, virtually wiped out by the northern pincer of the

PzKpfw 38(t) of 9.Kompanie, III Abteilung, Pz-Regt.21, 20.Pz-Div. photographed in Russia. The turret bears only the company number, in yellow; the divisional sign, a bar joining the top of three strokes, is in yellow on the rear engine access plate. This had been used in 1939–40 by 3.Pz-Div.—in an allusion to the Brandenburger Tor?—and would be dropped by 20.Pz-Div. in 1942. (US Nat.Archives)

PzKpfw 38(t) of Pz-Regt.204, 22.Pz-Div., which served in Russia only from February 1942 until it was virtually annihilated at Kalach that November, and was not re-formed. On the original the divisional sign—a yellow arrow pointing diagonally up and right, crossed by two short vertical strokes —is visible on the plate immediately below the exhaust. The turret number '522' has a bar beneath it, both being in white. Note that this crew have welded quite a large stowage rack of iron strip to the rear of the turret. (US Nat. Archives)

Soviet offensive on either side of Stalingrad in November 1942, Rumania's only armoured division had a single regiment of LTvz 35 tanks, some light CKD reconnaissance vehicles, the 3rd and 4th Motorised Rifle and 1st Motorised Artillery Regiments. The tank illustrated was finished in dark green with a random scrubbing of whitewash overall, mainly on the upper part of hull and turret. The national insignia, a double-pointed cross, was stencilled well forward on each hull side; and a two-digit tactical number appears to be in red trimmed with white.

G1: Panzerbefehlswagen 38(t), Stab Panzer-Regiment 25, 7.Panzer-Division; France, spring 1940

Finished all over in Panzer grey, this command tank version of the PzKpfw 38(t) has extra radio equipment and a large rear frame aerial. The red-and-white 'R02' identifies the deputy regimental command tank. Note that the narrow Balkenkreuz national marking is in white outline on the grey ground in the hull side positions, but in black, outlined white, on the centre rear of the turret bustle. In the latter position, below and to the right of the cross, appears the divisional sign— a reversed yellow Y-rune and three dots; this was probably repeated on the corner of the hull front plate by the driver's visor.

G2–G10: Panzer signal flags and pennants, 1940–42

Flags G2 to G5 are signal flags, whose meaning varied from time to time and according to the manner in which they were displayed. In the following list 'HU' means 'held upright from the turret', 'RL' means 'raised and lowered vertically', and 'W' means 'waved from side to side'. The usual size was about 12ins. × 16ins.; in some units plain coloured flags may have been replaced by batons.

G2: HU, 1940, 'Follow me'; 1941, 'Conform'; RL, 1940, 'Extended order'; W, 1940, 'Single file'. A blue flag meant: HU, 1940, 'Double file'; 1941, 'Extend'; RL, 1940, 'Close in'; W, 1940–41, 'Form arrowhead'. In 1940 a blue flag held down at a slant from the turret meant 'Open out'. A red flag meant: HU, 1940, 'Stand by'; 1941, 'Action' or 'Attack individually'; RL, 1940, 'Ready for action'; W, 1940, 'Attack artillery/AT/tanks'; 1941, 'Enemy tanks seen' or 'Attack in formation'.

G3: HU, 1940, 'Take up position'; 1941, 'Hull down'; W, 1940–41, 'Form broad arrowhead, base leading'. A yellow-over-blue flag meant: HU, 1940, 'Right turn'; 1941, 'Right wheel'; W, 1940, 'Left turn'; 1941, 'Left wheel'. A blue-over-red flag meant: HU, 1940, 'Take cover'; 1941, 'Turret down'; W, 1940–41, 'About turn' or 'Move'.

G4: In 1940–41, HU or W, this meant variations on the theme 'I am broken down/ a casualty' and 'I need assistance'. A yellow triangular pennant with a black capital 'W' (for 'workshop unit') was used for this purpose in Panzer-Regiment 8.

G5: 'Tank damaged' flag used early in 1942 by Pz-Regt.5.

G6: Probably tin plate, and used to mark location of HQ Pz-Regt. 5 in the desert; Pz-Regt.8 used the same but with red and black reversed; both may have had white backgrounds on occasion.

G7: In Africa both Pz-Regt.5 and 8 used this to mark Workshop Company locations.

G8: A tank company attacking British 2nd Armd. Div. units near Point 112 on 13/14 April 1942 were reported as all flying red flags. Large red flags were also used by German and Italian recce units, including light tanks, carrying out anti-LRDG/SAS sweeps in rear areas in August 1942.

G9: Reported flown by last tank in road columns in Africa, March 1942. Similar flag flown by lead tank in an attack by about 30 Panzers on 5th

RTR at Ruweisat Ridge, July 1942.

G10: Company commander's pennant, probably tin, used by Pz-Regt.5 or 8 in the desert. When cloth versions were used in conjunction with signal flags, this indicated that the order applied to the whole company. This pennant held upright meant 'All commanders to me'.

H1: PzBefw 38(t) Ausf.B command tank, 8.Panzer-Division; Russia, 1941

The yellow 'III' battalion headquarters marking, carried on turret and hull sides (see inset), probably identifies Stab, Panzer-Abteilung 67; this unit was, in all but name, the third battalion of the division's Panzer-Regiment 10. The divisional insignia, a Y-rune and a single stroke, is marked on the turret sides—most unusually—and also on the round plate blanking off the empty hull machine gun position of this command tank, which carried extra radio equipment instead. Later versions of the 38(t) command tank had a second rod aerial rising from midway along the left hand hull side in place of the frame 'bedstead' illustrated in G1; it is probably obscured by the crewmen in the photo on which we base this painting.

H2: LTvz 38, 11th Tank Co., Slovak Mobile Division; Russia, 1941

Initially the tanks of this unit were finished in the original Czech sand yellow, green and brown; later deliveries were in Panzer grey. The three-digit tactical number follows German style, but its exact meaning is unclear. The tank company operated LTvz35 and 38 tanks, and some PzKpfw IIs at a later date. The 10,000-man Mobile Division was an élite formation of two infantry and one artillery regiments and a recce battalion, all motorised, plus this tank company. It was employed alongside the best German units in the front line, and earned their ungrudging respect in the hard fighting of 1941–43. Nearly cut off at Krasnodar after covering the retreat from the Caucasus, it was very roughly handled, and morale suffered when the survivors were relegated to coastal defence, being reorganised as 1st Infantry Division (see D2). Details of the Slovakian and other German-allied contingents in Russia may be found in Men-at-Arms 131, *Germany's Eastern Front Allies 1941–45*.

A reminder that surviving but obsolete tanks, both German and foreign, were not wasted. Many were issued to training units, or second line units deployed on peaceful occupation duties in NW Europe. Others were stripped of their turrets and re-worked as SP guns of many kinds. On the Eastern Front many French, Czech, and other models were issued to armoured train units. Carried on flat-cars, they would disembark to give train commanders a mobile response when attacked by the partisan bands which swarmed in the rear areas. This PzKpfw 38(t), stripped of its hull machine gun, is finished in dull yellow with brown and green striping; markings seem to be limited to the black and white turret cross. (RAC Tank Museum)

Notes sur les planches en couleur

A1 Motif de camouflage gris et brun apparu pour la première fois en 1935. Le chiffre '5' semble identifier la Compagnie, et 'I', peint sur le parallélogramme blanc, indiquerait logiquement le Bataillon, bien qu'il n'y eût que quatre compagnies dans chaque bataillon. **A2** Unité mixte où des troupes Nationalistes espagnoles sont encadrées par des allemands. Le drapeau rouge et jaune des Nationalistes est peint en trois points différents; l'insigne blanc: mousquet, arbalète et halebarde est celui de la légion étrangère espagnole d'où provenait une partie des hommes; le losange blanc sur rouge désigne la 1ère Compagnie du 2ème Bataillon.

B1 'R02' indique le tank de commandement adjoint du régiment. Noter l'insigne divisionnaire à trois pointes peinte sur le bord arrière de la coupole du Commandant. La croix nationale a été obscurcie par de la boue ou de la peinture après que la 4. Panzer-Div a perdu une quarantaine de tanks au cours d'une bataille acharnée contre des canonniers polonais près de Mokra, le 1.9.39. **B2** 'IIL' indique que le 2e Bataillon portait l'épithète de 'léger'. Noter la nouvelle version de l'insigne divisionnaire.

C1 Le bison, insigne du régiment, a été appliqué au pistolet à peinture tout autour du bord d'un pochoir plein. Trait inhabituel: la tourelle ne porte que le numéro de la Compagnie, mais noter le numéro complet de la compagnie, troupe et le tank représenté sur la plaquette rhomboidale. Insigne divisionnaire à l'extrémité droite du blindage avant. **C2** Remarquer le point jaune suivant le numéro '216' sur la tourelle; il identifie le Pz. Reg 36 du Pz. Regt. 35 au sein de la division. Noter la large raie blanche sur le pont arrière servant de signe de reconnaissance aérienne.

D1 Exceptionnellement, ce char est en service dans un bataillon de reconnaissance et porte le repère: 'A 94' (A pour Aufklärung) à l'arrière de la tourelle et sur le côté du blindage devant l'insigne divisionnaire. **D2** Finition usine allemande jaune foncé uni, utilisée dès février 1943, avec l'écusson national de la Tchécoslovaquie sur les côtés de la tourelle. **D3** Tank de commandement du régiment à peinture jaune désert avec numéro particulier du tank '06' et l'insigne du régiment: un bison.

E1 Seules 30 de ces variantes appui lourd ont été réalisées et utilisées par une unité pilote servant en Russie dans la 1. Pz. Div. et plus tard en Yougoslavie. L'utilisation du motif 'L'ours de Berlin', trait habituel de la 3. Pz-Div. est inexpliquée mais confirmée par des photographies. **E2** Les modifications apportées à ce modèle de tank, avec matériel arrimé extérieurement, laissaient peu de place pour recevoir des inscriptions. Voici les codes à quatre chiffres des unités blindées de reconnaissance présentées en deux paires '41' et '34' peintes séparément sur une boite et une plaquette métallique fixée à un châssis de 'jerrycan' sur la tourelle.

F1 Finitions et inscriptions absolument standard pour l'époque; des petits numéros jaunes ont été peints sur les côtés et à l'arrière de la tourelle; et l'insigne divisionnaire sur les plaques de blindage arrière et avant. **F2** Insigne national sur le côté du blindage, bien à l'avant, et numéro de tourelle indiquant la sous-unité, quoique l'ordre ne soit pas bien connu.

G1 Fait rare, la croix et l'insigne divisionnaire sont peints sur la partie arrière de la tourelle de ce char de commandement. **G2–G10** Pour des raisons d'espace, il n'est pas possible de donner ici une traduction complète de ces longues légendes. Il y a donc lieu de se référer aux légendes en langue anglaise; noter que les abréviations utilisées sont: HU = tenu verticalement à partir de la tourelle; RL = Levé et baissé; W = flottant d'un côté à l'autre.

H1 Une deuxième antenne radio était montée sur le côté gauche, à mi-chemin le long du blindage., mais ici, elle est cachée par l'équipage. Noter le poste, obturé, de mitrailleuse, dans le blindage avant. Ici une radio supplémentaire remplace cette arme. **H2** Cette unité s'est acquise une bonne réputation en 1941–43. Le char bénéficie de la finition 'plan chèque' original: couleur sable, brun et vert, avec numéro de tourelle à trois chiffres à la manière allemande, mais la signification de ce numéro demeure obscure étant donné qu'il n'y avait qu'une seule compagnie blindée.

Farbtafeln

A1 Grau und braune Tarnung, eingeführt 1935, die Nummer '5' scheint die Kompanie zu bezeichnen, und die auf dem Parallelogramm aufgemalte 'I' würde dann das Bataillon identifizieren; abwohl es in jedem Bataillon nur vier Kompanien gab. **A2** Gemischte Einheit mit Truppen der spanischen Nationalisten neben deutschem Kader. Die rot-gelbe Flagge der Nationalisten ist dreimal aufgemalt; das weisse Abzeichen mit Muskete, Bogen und Hellebarde gehört zur spanischen Fremdenlegion, die ebenfalls Personal zur Verfügung stellte. Der Diamant (weiss auf rot) steht für die 1. Kompanie, 2. Bataillon.

B1 'R02' bezeichnet einen kleinen Panzerbefehlswagen. Man beachte das dreieckige Divisionsabzeichen auf der Rückseite der Kommandokuppel. Das Kreuz wurde mit Schmutz oder Farbe verdeckt, nachdem die 4. Panzerdivision in einer schweren Schlacht mit polnischen Artilleristen bei Mokra am 1.9.1939 etwa 40 Panzer verloren hatte. **B2** 'IIL' zeigt an, dass die 2. Abteilung ald 'leicht' eingestuft war. Man beachte die neuen Divisionsabzeichen.

C1 Der Büffel war das Regimentsabzeichen und wurde mit Hilfe einer festen Schablone aufgesprüht. Die Kuppel trägt gewöhnlich nur die Kompanienummer; in diesem Fall beachte man die volle Nummer der Kompanie, des Zugs und des individuellen Panzers auf einer rhomboiden Platte. **C2** Man beachte den gelben Punkt nach der Nummer '216' auf der Kuppel: dadurch unterschied sich das 36. vom 35. Panzerregiment innerhalb der Division. Man beachte ausserdem die breiten weissen Streifen auf dem hinteren Deck, ein Lufterkennungszeichen.

D1 Dieser Panzer wurde ausnahmsweise in einer Aufklärungs-Abteilung eingesetzt und trägt die Nummer 'A (= Aufklärung) 94' auf der Hinterseite der Kuppel sowie vorne vor dem Divisionszeichen auf der Karosserie. **D2** Einfacher dunkelgelber Deckanstrich, wie deutsche Fabriken ihn seit Februar 1943 produzierten, mit dem Nationalwappen der Slowakei auf den Kuppelseiten. **D3** Regiments-Stab-Panzer, mit wüstengelber Bemalung, individueller Panzernummer '06' und dem Büffel-Abzeichen des Regiments.

E1 Nur 30 dieser schweren Hilfspanzer wurden gebaut und mit Versuchseinheiten der 1. Panzerdivision in Russland und später auch in Jugoslawien eingesetzt. Für den Berliner Bär, der gewöhnlich von der 3. Panzerdivision getragen wurde, gibt es keine Erklärung; die Verwendung ist allerdings durch Fotos bestätigt. **E2** Die zahlreichen Zusätze und aussen angebrachten Ladevorrichtungen lassen wenig Raum für Markierungen; bei diesem Modell sind die von Aufklärungseinheiten benutzten vierstelligen Codeziffern in zwei Paaren, '41' und '34', auf einen Kasten und eine Metallplatte aufgemalt, die an einem Benzinkanister-Gestell auf der Kuppel angebracht sind.

F1 Für die Zeit üblicher Deckanstrich mit Standard-Markierungen; kleine gelbe Zahlen wurden hinten und an den Seiten auf die Kuppel gemalt, die Divisionszeichen stehen hinten und vorne auf Karosserieplatten. **F2** Landesabzeichen weit vorne auf der Karosserie; eine Zahl auf der Kuppel identifiziert die Untereinheit (Reihenfolge nicht erklärt).

G1 Kreuz und Divisionszeichen sind ausnahmsweise bei diesem Panzerbefehlswagen hinten auf die Kuppel gemalt. **G2–G10** Aus Platzgründen kann keine Übersetzung der langen Bildunterschriften gegeben werden. Es wird auf die englischen Artikel verwiesen; man beachte die Bedeutung der Abkürzungen HU (senkrecht von der Kuppel aus hochgehoben), RL (senkrecht erhoben und gesenkt) und W (seitwärts hin und her bewegt).

H1 Etwa in der Mitte der Karosserie war eine zweite Radioantenne angebracht, hier von einem Besatzungsmitglied verdeckt. Man beachte das abmontierte MG vorne auf der Karosserie, stattdessen wurden zusätzliche Funkgeräte getragen. **H2** Diese Einheit genoss einen guten Ruf in den Jahren 1941–43. Der Panzer hat den üblichen tschechischen Deckanstrich in sandfarben, braun und grün mit einer dreistelligen Kuppelnummer im deutschen Stil; da es nur eine Panzerkompanie gab, ist die Bedeutung unklar.